D0782272

DATE DUE

JAN 1 3 2013	

BRODART, CO.   Cat. No. 23-221

Romancing the Atom

Military and civilian VIPs wearing protective goggles watch an atmospheric test from the officers' beach club on Parry Island, Enewetak Atoll, April 8, 1951. The test, called Greenhouse Dog, is 12.5 miles away on Runit Island. (Defense Threat Reduction Agency (DTRA))

Romancing the Atom

Nuclear Infatuation from the Radium Girls to Fukushima

ROBERT R. JOHNSON

 PRAEGER

AN IMPRINT OF ABC-CLIO, LLC
Santa Barbara, California • Denver, Colorado • Oxford, England

Library of Congress Cataloging-in-Publication Data

Johnson, Robert R., 1951–
  Romancing the atom : nuclear infatuation from the radium girls to Fukushima / Robert R. Johnson.
     p. cm.
  Includes bibliographical references and index.
  ISBN 978–0–313–39279–5 (hbk. : alk. paper) — ISBN 978–0–313–39280–1 (ebook)
1. Nuclear industry—Social aspects—United States—History. 2. Atomic bomb—Social aspects—United States—History—20th century. 3. Nuclear industry—Health aspects—United States. I. Title. II. Title: Nuclear infatuation from the radium girls to Fukushima.
HD9698.U52.J65   2012
338.4'7621480973—dc23   2012011724

ISBN: 978–0–313–39279–5
EISBN: 978–0–313–39280–1

16 15 14 13 12   1 2 3 4 5

This book is also available on the World Wide Web as an eBook.
Visit www.abc-clio.com for details.

Praeger
An Imprint of ABC-CLIO, LLC

ABC-CLIO, LLC
130 Cremona Drive, P.O. Box 1911
Santa Barbara, California 93116-1911

This book is printed on acid-free paper ∞

Manufactured in the United States of America

For our twin granddaughters, Katelyn and Sydney

and

To the memory of Paul S. Boyer (1935–2012)

"Henry Adams once wrote: 'No honest historian can take part with—or against—the forces he has to study. To him, even the extinction of the human race should merely be a fact to be grouped with other vital statistics.' I readily confess I have not achieved Adams's austere standard of professional objectivity. This book is a product of experiences outside the library as well as inside, and it is not the work of a person who can view the prospect of human extinction with scholarly detachment."

—Paul S. Boyer, Introduction to *By the Bomb's Early Light: American Thought and Culture at the Dawn of the Atomic Age*

Contents

Acknowledgments

As any author will attest, there are always many people to thank for the development of a book, and so it has been with this book over the last four years. However, the seeds for this project were planted long before I had put any words to paper. In the summer of 1993, I was teaching a technical writing course at Miami University. That was the summer following the discovery of the highly secret Alba Craft, Inc., uranium milling site in a residential area of Oxford, Ohio. In that class, the students conducted archival research, conducted interviews with men who had worked in the facility in the 1950s, and wrote reports and articles on the recently discovered radioactive site for the student newspaper and local press. So it is fitting that I begin by thanking the members of that technical writing class.

The seeds for this book remained dormant in my mind, however, for about fifteen years until I had a conversation with my sisters, Judy Shubert and Carol Pfefferkorn, and my wife, Evie, in Williamsburg, Virginia. I was contemplating a book project, but not on the topic that would become this book. For some reason that I cannot recall, I told them the story of the Oxford, Ohio, secret uranium milling site, when suddenly, in unison, they said, "Do your book on that. It is such a good story!" So I did, and I thank them for their early enthusiasm and continued support by passing along newspaper and magazine articles and providing commentary on early drafts of chapters.

I also want to thank my brother and sister-in-law, H. Thomas Johnson and Elaine B. Johnson, for their enthusiastic support. They are both authors and understand the importance of cheering another writer on. And my brother- and sister-in-law, Peter and Donna

Svirsky, who provided both enthusiasm for the project as well as miscellaneous news articles they thought I might be able to use.

My editor at Praeger, Michael Millman, deserves my highest praise for his honest and very helpful feedback and editing. He has believed in this project from the beginning and has helped to make it much better through his expertise and consistent support.

I could not have gotten such a good start on the manuscript without the time afforded by a sabbatical in the fall of 2009. For this, I thank the interim chair of my department, Heidi Bostic, and the Research Office at Michigan Tech for providing scholarship funds that supported my travel for interviews and off-campus research. I am indebted to Heidi Julavits and Andrew Leland of McSweeney's Press, who edited an earlier version of the chapter on the Radium Girls that appeared in the September 2010 edition of *The Believer*. Caitrin Nicol, managing editor of *The New Atlantis: A Journal of Technology and Society*, has my gratitude for editing an early version of Chapter Two that appeared in the Summer 2009 edition.

The assistance I received from the library at Michigan Tech made my archival research much easier. Every author should be lucky enough to draw upon the expertise of people such as Nora Alred, Mies Martin, and Brianna Williams when it comes to finding those ever-elusive historical documents that are hidden away in the new digital archives and in those good old-fashioned microfilm collections. I also received very important assistance from Peter Kuran of VCE, Inc. who helped me procure several of the historical photographs that are included in the book.

The conversations and interviews I had with Paul Boyer, Linda Musmeci Kimball, Jack Niedenthal, Yerevan Peterson, Mike Mullins, Harvey Wasserman, and Eileen Welsome over the past three years provided so much invaluable knowledge that I cannot begin to thank them enough. I give an especially warm thank you and debt of gratitude to Linda and Yerevan for providing the archival materials of the Oxford Citizens for Peace and Justice and their own personal records, and to Jeff Kimball—a highly regarded historian of the atomic age—for the conversations we had at their home when he and Linda so graciously put me up (and put up with me) on several trips to Oxford.

Finally, I give my deepest gratitude to my wife, Evie. She is an unparalleled editor, patient listener, insightful commenter, exquisite collaborator, and something she probably never thought she would be: an expert in Chicago Style! She "touched" every page of this book. Including this one.

Introduction

The release of atomic power has changed everything except our way of thinking ... the solution to this problem lies in the heart of mankind. If I had only known, I should have become a watchmaker.

—Albert Einstein[1]

If you are in residence on Earth, even if you are more than 100 years old, you have been romanced by the atom. You might be a resident of the so-called developed world, or you might be a member of an indigenous community anywhere on this planet. In fact, if you live in a part of the so-called developing world, you might have been more affected by this romance than if you lived in New York, Paris, Moscow, Sydney, Buenos Aries, Mexico City, or any other major metropolitan area of the world. You could make your living from the land, from the water, from a factory, from an office—no occupation or way of life is exempt from our powerful infatuation with the atom. You may consider yourself a victim, a beneficiary, or an innocent bystander of this romance. Whatever the case, we are all held in some fashion to this common bond of infatuation with the atom.

This common bond, however, is a paradox. On one side, the bond creates a togetherness, a community of believers. Albert Einstein is but one example of a leader of the atomic development community. He created theories of the atom that revolutionized the discipline of physics and virtually all of modern science, and he also was a prominent player in making things with the atom. Most notable, of course, was the atomic bomb.

Einstein came to realize, too, that the common bond had a flip side: one that makes the bond a constraining bind, and not only for atomic developers, but for the whole of humankind. As he lamented less than a year after the United States dropped two atomic bombs on Japan, "The atomic bomb has changed everything except our way of thinking." One of the greatest thinkers of all time, Einstein fully understood the immense power of thinking. The powerful force of thinking had met its match, however, with the advent of the atomic bomb. And he also believed that thinking could possibly overcome this entrapping bind of the human mind, if only humans could think in different ways.

Another great thinker writing at the same time as Einstein—the philosopher Martin Heidegger—also advocated for other ways of thinking. He described human thinking in the modern era as being of two types: calculative and meditative. Calculative thinking, which Heidegger holds up as the dominant mode of thinking in the twentieth century, "races from one prospect to the next, ... never stops, never collects itself."[2] Calculative thinking thus is a straight-ahead way of thinking that eschews alternative possibilities because it is firmly set in its mission to accomplish a specific goal, such as making an atomic bomb.

Alternatively, meditative thinking "demands of us not to cling one-sidedly to a single idea, not to run down a one-track course of ideas. Meditative thinking demands of us that we engage ourselves with what at first sight does not go together at all."[3] Einstein may not have been familiar with Heidegger's concept of calculative and meditative thinking, but I believe he would have found some truth in the distinction, as it potentially defines different ways of thinking that Einstein felt mankind was lacking. Further, instead of being represented as a simple binary, if calculative and meditative thinking are laid out across a spectrum ranging from "highly calculative" to "highly meditative," then there are even more possibilities to define other ways to think.

A central argument of this book is that humans have frequently been operating from the extreme end of the calculative thinking spectrum when it comes to our infatuation with the atom. Put directly, we have become locked into a *mindset*: an unreflective set of beliefs that we construct about something we might hold dear, something we might fear, something we cannot live with or without. This mode of thinking is literally, as the word makes explicit, "set." In *Nukespeak: Nuclear Language, Visions, and Mindset* (1982), Stephen Hilgartner,

Richard C. Bell, and Rory O'Connor explain that, "The word mindset means what it implies, a mind that is already set. A mindset acts like a filter, sorting information and perceptions, allowing some to be processed and some to be ignored, consciously or unconsciously."[4]

Further, *Romancing the Atom* will repeatedly demonstrate how the atomic mindset is created through humans' myriad *uses* of the atom and the immense curiosity and power that atomic usage has held over humans in the last hundred-odd years to become an obsession of human consciousness and, ultimately, action. After all, the atom is an element of nature and, thus, is essentially innocent of any human actions until humans decide to put it into use. Once fostered into use by humans for specific purposes, the atom becomes a technology—a tool to be manipulated as humans see fit. Humans ultimately have the ability to make choices about the uses of the atom, but the calculative thinking of the mindset often moves quickly beyond possible alternatives and, as Heidegger (1966) suggests, "never stops, never collects itself."[5]

Mindsets can be further complicated due to their subtlety. They operate through belief systems that are as often as not unexamined, and they can sneak up on us. Once mindsets are formed, we are supposed to believe in them. Beyond the atomic mindset, humans are surrounded by a wide variety of mindsets that operate through various venues. One example is the mindset of free enterprise economics. The idea of unlimited free markets caught the imagination (and pocketbooks) of people worldwide during the past several decades. Profits were made hand over fist—until, of course, the markets could no longer sustain the unexamined enterprise. A belief in unlimited growth drove these markets up and up until growth slowed and nearly stopped. The mindset of unlimited growth that would never end created a belief system that moved forward and took on a life of its own. We even personified markets as human. For example, we were often told—in one form or another by the government, banks, or media—that financial markets act rationally or irrationally, or that the markets "must be convinced" that the economy is stable enough for investments to happen. Giving agency and human thought capabilities to commodity and financial markets provided the illusion that the markets are living, breathing things and we can do little to affect their development, as they will grow and develop on their own. Using a metaphor of concrete, a mindset of free market systems set the notion of free markets in our minds as all for the good, period. Our minds became hardened and, in turn, we continued believing that the markets themselves were in charge. In reality, mindsets are always created and manipulated by humans, who make decisions based upon these unreflective beliefs.

Mindsets are also about forgetting. When the wet concrete of mindset belief systems is set, humans have a tendency to forget what fosters them and what consequences they have created over time. Opening mindsets to examination aids in remembering, and remembering is often best told through stories. The complications of the atomic mindset and issues of control, secrecy, naiveté, and many other human qualities will be central to each story in this book. Thus, each chapter will tell a story—some simple, some more complex. But each depicts moments in the twentieth and twenty-first century of this infatuation with the atom and its cultural, social, political, individual, and environmental consequences. To begin this journey into atomic infatuation, I offer two brief examples of the atomic mindset that span the twentieth century.

Dr. William J. Morton[6] was quite the character: Harvard-educated physician, technology inventor, entrepreneur, marketing expert, and speculator, among other things. Looking into his life that spanned the last half of the nineteenth and first quarter of the twentieth centuries, you will find that he is most often remembered for two things. First was a moment of fame he achieved when he spent six months in jail with Julian Hawthorne (the son of the American writer Nathaniel Hawthorne) for mail fraud violations pertaining to selling interests in Canadian mining properties. His other moment of fame derives from his infatuation with radioactivity. Morton was credited with developing various medicines and instruments that could be used by physicians in the supposed treatment of a variety of human ailments. Chief among his "discoveries" was what he called "Liquid Sunshine." Liquid Sunshine was a radioactive elixir concocted from radium that was drunk by patients to put "sunshine" into the interior of bodies so that the inner organs could reap the same benefits from sunshine afforded to the outer body by the sun itself.

Morton had a number of followers, as the use of radiation was a growing prospect for his contemporary medical community. Those interested in his products included a number of quacks and hacks but also some of the most prominent medical practitioners and researchers from across the globe. His fame had grown to the point that in 1904, the alumni association of the Massachusetts Institute of Technology, the Technology Club, held a dinner in his honor. At the dinner there was a cavalcade of radioactive displays that included human skeletons covered in luminous paint, balloons painted with luminous paint, and pasteboard chickens similarly decorated. In addition, at the table of each guest were small glasses of Liquid Sunshine. At the conclusion of the dinner, with the lights dimmed to display the luminous quality of the liquid, Morton raised a toast of the mysterious elixir and, "amid great applause and enthusiasm," guests drank the liquid like a shot of whiskey.

As we shall see in subsequent stories, the fascination with radium as a curative was widespread and would hold the imagination for many years to come. Morton was literally, in many ways, a radioactive pioneer.

David Hahn was a shy boy growing up in suburban Detroit in the 1980s when he became very interested in all things radioactive.[7] He was given a chemistry set at age twelve and, in a few short months, had learned how to make nitroglycerin. He subsequently caused an explosion in his parents' basement that forced a ban on such experiments in the house. So David found another space in the backyard: a tool shed. Undeterred, David's fascination turned to bigger things, things atomic. To further his foray into atomic elements, David wrote a letter to a director of the Nuclear Regulatory Commission (NRC) inquiring about the presence of radioactive elements in common, everyday things. Encouraged by a quite helpful response from the NRC director—he was told that the amounts of such materials are insufficient to be of risk—David set about collecting materials from such mundane objects as the mantles of gas lanterns (these contain thorium 232) and the batteries of used smoke detectors (these contain americium 241) and was lucky enough to find a small vial of radium paint inside an old clock that he bought at an antique store for $10.

By the time David turned 17, he had experimented in numerous ways with the various elements he had collected. But he wasn't very satisfied with his work, as he hadn't really accomplished his larger goal: creating a small nuclear reactor. After studying some high school and college textbooks, David combined products from his radioactive cache and created an atomic cocktail of radium and americium with beryllium and aluminum. He wrapped it in aluminum foil and created, in essence, a rudimentary reactor core. "It was radioactive as hell," David recounted.[8] His Geiger counter had picked up elevated levels of radiation five houses down from his parents' home. David realized that maybe he was playing with something that was too dangerous "all in one place," so he divided the material up, storing some in his parents' house, some in the shed, and some in the trunk of his car.

On the night of August 31, 1994, local police were summoned by a neighbor who believed someone was stealing car tires. Police arrived to find David, who said he was waiting for a friend. Suspicious of David and the car, the police searched the trunk, where they found a locked toolbox sealed with duct tape. Inside they found some odd gray powder. David told them that it was radioactive. Believing that they might have discovered an atomic bomb, the police called in the Emergency Response Team.

For several weeks, workers dressed in radioactive-protection suits scoured his parents' house, yard, and shed and the entire neighborhood. In the end, they filled 39 steel drums with radioactive debris to be shipped to a disposal site in Utah. "I wanted to make a scratch in life," David said after the episode was over. "I don't believe I took more than five years off my life."[9]

NOTES

1. *New York Times*, "Bikini Climax," 1946. Article includes quotes from Einstein's letter pleading for a "new way of thinking."

2. M. Heidegger, *Memorial Address*, 1966, 44.

3. Ibid., 44.

4. S. Hilgartner, *Nukespeak*, 1982, xiii.

5. M. Heidegger, *Memorial Address*, 1966, 44.

6. S. Hilgartner, *Nukespeak*, 1982, 5.

7. K. Silverstein, *Boy Scout*, 2005. The story of David Hahn is drawn from this book.

8. Ibid., 137.

9. Ibid., 192–193.

PART ONE

Inventing the Atomic Mindset: Dial Painters, Comic Books, and Paradise Lost

Everything has to start somewhere, and so it is with the atomic mindset. Beginnings are called by numerous names—origins, starting points, openings—but one term that calls forward the notion of beginning is *invention*. Derived from the ancient Greek and Roman term *inventio*, invention has two specific meanings. First is the notion of "making new." This is the most commonly used sense of the term in our modern age and, in fact, has come to be a driving concept in virtually all contemporary technologies. We continually invent, reinvent, make, remake, and minister to the development of new technologies faster and more completely than just about any other human activities. In the realm of things atomic, the inventing of technologies that use the atom as a starting point has been an all-consuming activity around the globe and has been no less than tantamount to the making of worlds. Or, as the nuclear physicist Robert Oppenheimer proclaimed in 1945 by quoting from the Bhagavad Vita shortly after he witnessed the first atomic explosion in the New Mexico desert, "I am become Death, the destroyer of worlds." Invention has its consequences, and some of them are not always apparent at the beginning, as we see repeatedly in the atomic mindset.

The second concept derived from *inventio* is inventory: a collection of things both material and linguistic that are used as an aid in making. In the making of a mindset, there are countless inventories of artifacts, results from experiments, and theories that inform and depict the mindset. *Romancing the Atom* plays a role in this inventory of things atomic through a collection of stories that begins with early moments in the invention of the romance and mindset that emerged in roughly the first half of the twentieth century.

This initial section provides three stories of how the atomic mindset and its romance with the atom began. First is the unfolding of the events surrounding the plight of the radium dial painters during and immediately following World War I. The chapter concentrates on four women known in the press as "the radium girls." These women were representative of thousands of young women who used a radio-active luminescent paint, patented under the name Undark, to make the numbers of watch and clock dials (among other things) visible in the dark. Unknown to them, the paint was a highly toxic concoction that would eventually result in their demise after protracted and con-tentious legal battles, battles that resulted in significant changes to the rights of industrial workers.

Chapter 2 continues the invention of the mindset through two events. It begins with the pre–World War II communications among Albert Einstein, President Roosevelt, and a couple other physicists that set the stage for the development of the first atomic bombs. Fol-lowing the war and the dropping of the two bombs on Hiroshima and Nagasaki, Japan, in 1945, the story of the frenzied hunt for ura-nium in the American Southwest depicts the intense, and sometimes lucrative, exploration for uranium that created a boom atmosphere that took the entire nation by storm. Concomitant with the uranium boom was a euphoric love affair with the atom that brought the whole atomic enterprise to the popular culture of the country through such products as board games, atomic science kits for budding young scien-tists, and even comic books that used popular characters to foster the atomic imagination of the American public.

The final chapter in this section begins to make the bridge from the inventing of the atomic mindset to the using of the mindset through the story of the Bikini Islanders of the South Pacific. In 1946, the American government, mostly represented here by military and scien-tific interests, wanted to display to the rest of the world that they had control of this newfound ultimate weapon. To make the demonstra-tion possible, they convinced the residents of the Bikini Islands to allow them to use their land as a testing site for atomic bombs. The Bikinians were offered little in this bargain, save that the islanders were told these tests would be "for the good of mankind." The story follows the military and scientific experimentation exploits, and the Bikininan people, through a long and never-ending odyssey of nearly mythic proportions.

CHAPTER 1

For the Love of Pretty Things: The Radium Girls and "Dying for Science"

They say if you visit the grave of Katherine Schaub in Newark, New Jersey's Holy Sepulcher Cemetery and bring a Geiger counter, the machine will register a significant positive reading. The source of the radiation is in the bones of Katherine. She was buried in 1933 after more than a decade of excruciating suffering that was the result of the decay of her bones, teeth, jaw, skull, and brain: a disease chillingly named *radium necrosis*.

Katherine was a Radium Girl. This is her story and that of her four fellow radium dial painters—Edna, Grace, Quinta, and Albina—and hundreds who remain unnamed: a story of deceit, corruption, denial, and maybe above all, innocence.

In 1902, William Hammer, an American engineer and inventor, arrived in the United States with a gift given him by Pierre and Marie Curie: radium salt crystals. Hammer was fascinated by the potential attributes of his new gift. Not as interested in the promise of medical cures that had made radium such a hot topic at that time, Hammer instead was enamored by the intense radiance that the newly discovered element provided. He experimented with various concoctions and eventually found that a combination of glue, zinc sulfide, and radium crystals would create an iridescent paint. The paint could be applied to just about anything to produce a glow-in-the-dark effect: wristwatches and clocks, gun sights, and ornaments for display on houses and even on human bodies. Fingernail polish was a popular use for the new substance. Children's toys were not exempt: dolls' eyes glowed, and toy trucks and cars sported glow-in-the-dark lights and dials. There was a strong demand for luminescent timepieces and other radium-painted objects at this time, reaching the level of fad by the 1920s.

Hammer sold the rights for his radioactive concoction in 1914 to a New Jersey company, owned by Dr. Sabin Arnold von Sochocky and Dr. George S. Willis, named the Radium Luminous Material Corporation. They eventually changed the name to U.S. Radium Corporation in 1921. In 1917, they set up shop in Orange, New Jersey, intent on developing a market for their product, and they assured the public that the radium was in "such minute quantities that it is absolutely harmless." They dubbed the paint-like substance Undark.

Grace Fryer, Edna Hussmann, Katherine Schaub, and sisters Quinta McDonald and Albina Larice were all hired in about 1917. They ranged from 15 to 19 years of age at the time of hire and were five of the more than 2,000 dial painters employed by U.S. Radium over the next couple of decades. Their job was straightforward: to paint the dials of watches, clocks, and military instrument panels for use in ships, airplanes, and other equipment that demanded nighttime use. Most of the time, however, they just painted watches. They were proud of their work, especially since it supported the war effort. "I was pleased with the idea of a job which would engage me in war work. Some of the young women would scratch their names and addresses into these watches, and sometimes a lonely soldier would respond with a letter," Katherine said.[1]

The work was tedious, and the conditions were less than ideal: It was hard on the eyes, back, and hands, due to close work and repetitive actions while sitting at wooden benches and tables. Nevertheless, the work was steady and it paid better than what most women, especially first-generation Americans, could get following the return of servicemen from World War I. They were paid by the piece—one and a half cents per watch—and many women made $20 per week by painting about 250 watches a day over a 50-hour week. This compared to the median income of women workers in New Jersey of about $15 per week. These were considered to be good jobs for working women.[2]

The work was also exacting. The numbers and symbols on the dials were often small, and the camelhair brushes had to be kept sharp. To accomplish this, the women were instructed by their managers to simply lick the ends of the brushes often. This would keep them nice and pointed, ready for the meticulous work. Undark, after all, was tasteless and odorless. "I think I pointed mine with my lips about six times to

The Power of Radium at Your Disposal

Twenty-three years ago radium was unknown. Today, thanks to constant laboratory work, the power of this most unusual of elements is at your disposal. Through the medium of Undark, radium serves you safely and surely.

Does Undark really contain radium? Most assuredly. It is radium, combined in exactly the proper manner with zinc sulphide, which gives Undark its ability to shine *continuously* in the dark.

Manufacturers have been quick to recognize the value of Undark. They apply it to the dials of watches and clocks, to electric push buttons, to the buckles of bed room slippers, to house numbers, flashlights, compasses, gasoline gauges, autometers and many other articles which you frequently wish to see in the dark.

The next time you fumble for a lighting switch, bark your shins on furniture, wonder vainly what time it is *because of the dark*—remember Undark. *It shines in the dark*. Dealers can supply you with Undarked articles.

For interesting little folder telling of the production of radium and the uses of Undark address

RADIUM LUMINOUS MATERIAL CORPORATION
58 PINE STREET - - - - NEW YORK CITY
Factories: Orange, N. J.   Mines: Colorado and Utah

To Manufacturers

The number of manufactured articles to which Undark will add increased usefulness is manifold. From a sales standpoint, it has many obvious advantages. We gladly answer inquiries from manufacturers and, when it seems advisable, will carry on experimental work for them. Undark may be applied either at your plant, or at our own.

The application of Undark is simple. It is furnished as a powder, which is mixed with an adhesive. The paste thus formed is painted on with a brush. It adheres firmly to any surface.

UNDARK
Radium Luminous Material
Shines in the Dark

"The Power of Radium at Your Disposal." Advertisement for Undark, circa 1921. (Archived at Argonne National Labs.)

every watch dial. It didn't taste funny. It didn't have any taste, and I didn't know that it was harmful," Grace would later say.[3]

Drs. von Sochocky and Willis, as well as the scientists who worked for U.S. Radium producing Undark, knew better. The key ingredient

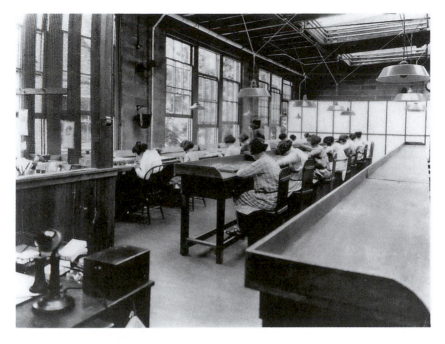

Radium dial painters at work in East Orange, NJ, circa 1921. (Archived at Argonne National Labs.)

of Undark is about one million times more radioactive than uranium. Even though the scientists didn't know about all of the radioactive dangers that would be revealed in the decades to come, they knew very well that the substance was hazardous. Indeed, the scientists protected themselves with lead shields and used masks and tongs to handle Undark production processes. None of this danger was known outside the science lab. It was, like so many secrets of the atomic mindset, kept under lock and key.

In 1920, Grace was offered a more appealing position as a bank teller. She liked to be in contact with people, so she took the opportunity. Grace left U.S. Radium to improve her situation. Tragically, U.S. Radium would never leave her.

About two years into her bank job, Grace started to feel ill. She didn't have her usual youthful spunk, and she was experiencing various pains in joints, particularly in her mouth and jaw. The symptoms became so extreme that when her teeth began falling out and a painful abscess formed in her jaw, she sought medical attention from several doctors. Frustratingly, they could offer no relief because they had never seen anything like her malady.[4] Some doctors even told her that she looked healthy, as her skin had a nice rosy hue. The radium, unknown to the doctors, was causing a temporary increase of red blood cells, something that is common even today in cancer patients undergoing nuclear medicine treatments. The radium in Grace's body, though, was unchecked and in such enormous quantities that the cause of her rosy glow would soon enter her bone marrow and turn her skin from the hue of a garden flower to that of fireplace ash.

Grace continued to seek medical attention for three years, until 1925, when she was referred by a friend to a Columbia University specialist—Frederick Flynn. He examined her and declared that she was in fine health. A colleague present at the exam concurred with Flynn's diagnosis. It would be some time before Grace learned that Flynn was not a licensed medical doctor; he was an industrial toxicologist under contract with U.S. Radium. His colleague was also not a doctor: He was the vice president of U.S. Radium. This connection with U.S. Radium and the two "doctors"—something that would not be revealed until court hearings several years later—was just the tip of an iceberg of deceit perpetrated by U.S. Radium to conceal the hazards of Undark.

About the same time that Grace was seeking medical assistance, U.S. Radium had contracted with a noted Harvard toxicologist, Dr. Cecil Drinker, to conduct a study of working conditions at U.S. Radium's New Jersey facilities. Drinker was a highly respected scientist who, at the time of the U.S. Radium operation, was helping to develop the field of industrial hygiene. He had begun a research facility at Harvard in the School of Public Health and had studied the effects of industrial dust on respiration and blood content of workers in the zinc industry. He eventually concluded that the culprit in these instances of industrial poisoning was manganese. U.S. Radium was his first foray into studying the industrial hazards of radiation, a foray that would last for nearly a decade and define Drinker as a primary pioneer of the field.

Drinker examined the workplace in Orange and observed an environment replete with radium-tainted dust, open containers of highly radioactive paints, poor ventilation, and other problematic conditions. He also took blood samples from the workers on the shop floor as well as the scientists working in the adjoining labs. What he found was disastrous. Every one of the workers had dangerous blood conditions that were not easy to explain. He also encountered several cases of radium necrosis, the disease that would eventually take the lives of the Radium Girls and many coworkers. He noticed, too, that a chemist, Edward Lehman, had severe lesions on his hands and arms. Lehman dismissed the idea that Undark had anything to do with his lesions or that there was any threat to his future health from the substance. Lehman would die within the year.

Drinker remarked that Lehman's attitude of complacency was rampant at the company. "There seemed to be an utter lack of realization of the dangers inherent in the material that was being manufactured."[5] Such complacency is only to be expected in employees of a company that was led by someone who saw absolutely no problem with radium products. U.S. Radium even sold the sand-like residue of the radium paint process as filler for children's sandboxes. When some parents questioned the safety of the sand, von Sochocky assuaged them by telling them that the sand was "most hygienic and ... more beneficial than the mud of world renowned curative baths."[6] Denial, coupled with deception and fraud, were becoming markers of the developing atomic mindset.[7]

In June 1924, Drinker wrote an extensive report following his investigation and presented it to Arthur Roeder, now the president of U.S. Radium. (Dr. von Sochocky was no longer in charge, as he had sold the business to Roeder and was fighting his own battle with radium poisoning. He would die about the same time as the Radium Girls' trial.) Roeder responded in writing to Drinker and claimed that there were problems with the findings of the report and that he would provide facts to demonstrate the misinterpretations.

Roeder never responded to Drinker, thus stalling the completion of the report. He also forbade Drinker to publish the report in a scientific journal because he claimed that Drinker had agreed to complete confidentiality about his research at U.S. Radium. Drinker never did have such a confidentiality agreement, but this stalling technique gave Roeder enough time to carefully edit the report before releasing it to the New Jersey Department of Labor. Drinker knew nothing of Roeder's editing until, through the help of a colleague, he was informed of the startling changes to the report.

In April 1925, Dr. Alice Hamilton—the first female professor at Harvard—was on the board of the Consumers League, an organization formed in 1899 that worked to protect workers' rights.[8] Hamilton had received a copy of Roeder's altered report by the New Jersey Labor Board. Now in the hands of Hamilton, the report had an informed reader who could understand the insides of the story. Hamilton, immediately upon seeing Roeder's report, wrote a letter to Katherine Drinker (Cecil's wife, also a PhD and member of the Harvard research group) from Hull House in Chicago where she was in residence. In the letter, she said, "Mr. Roeder is not giving you and Dr. Drinker a very square deal ... The Labor Board has a copy of the report and it shows that 'every girl is in perfect condition.' Do you suppose Roeder could do such a thing as to issue a forged report in your name?"[9]

Roeder's "editing" and forging of his name was the last straw for Cecil Drinker. Although threatened by Roeder and U.S. Radium, he published the report in the August 1925 issue of the *Journal of Industrial Hygiene* under the title "Necrosis of the Jaw in Radium Workers." The ghastly vision of the radium workers' bodies as described in the report was vastly different from what Roeder had reported to the Labor Board: "Their hair, faces, hands, arms, necks, dresses ... were luminous. One of the girls showed luminous spots on her legs and thighs. The back of another was luminous almost to the waist."[10] The Radium Girls' condition clearly was not so "perfect." The revelation of the actual conditions and hazards of the U.S. Radium shop would now set off a flurry of events that would eventually help to shape the field of industrial hygiene and worker compensation law, but the road would have many obstacles along the way, especially for the Radium Girls.

Using her connections through the Consumer League, Hamilton telephoned Walter Lippmann, editor of the influential *New York World*, a newspaper founded by Joseph Pulitzer and known for reporting on the plight of workers and low-income groups. Together, they began to develop a plan for disseminating stories concerning the Radium Girls and U.S. Radium.[11]

Grace had already seen many doctors by 1926, and the results were everything from (falsely) positive, as with Frederick Flynn, to those of at least one doctor who believed that her condition was related to her work at U.S. Radium. Her health deteriorated by the day, and with little hope coming from the doctors, Grace turned to the legal profession. This task, just like the medical investigations, became problematic and frustrating. Grace consulted a number of lawyers in New Jersey and New York, but to no avail. It wasn't that they were not interested in what could be a lucrative lawsuit. Rather, U.S. Radium had many friends in high places, and these lawyers felt that it wasn't worth the professional risk to help a poor working-class woman. Grace was persistent and finally, after two years, secured a young attorney from Newark, Raymond Berry, who filed a suit on her behalf in May 1927: *Grace Fryer vs. U.S. Radium*. Edna Hussmann, Katherine Schaub, and sisters Quinta McDonald and Albina Larice joined in the lawsuit and sought $250,000 in compensation plus medical expenses already incurred.

The legal process would prove to be protracted, beginning with the problem that the State of New Jersey had a two-year statute of limitations on workplace injury claims. Berry argued that the statute applied to the time the victims learned of their affliction, not from when they had left U.S. Radium more than two years prior to the lawsuit. He managed to get a judge to rule that the case should be brought to a hearing by presenting it to the New Jersey Court of Chancery, a court that harkens back to fourteenth-century England and has been used in U.S. courts to rule on legal disputes concerning issues of equity that are not covered under existing laws. As King James explained of the role of chancery in the seventeenth century: "Where the rigor of the law in many cases will undo a subject, then the chancery tempers the law with equity, and so mixes mercy with justice, as it preserves a man from destruction."[12] Thus, with a court that mixed mercy with justice, Berry was now ready to bring the case to court based upon the chancery decree.

Then, quite apart from Berry's work, two things happened to aid the Radium Girls' case.

A former dial painter who also happened to be a sister to Quinta and Albina, Amelia Maggia, had died in 1922, a death attributed to syphilis. A New York dentist, Joseph P. Knef, had treated Amelia and grew suspicious. He had removed part of Amelia's jaw only months before her death. The jaw was full of rot and holes that looked like necrosis caused by phosphorus. The diagnosis that was termed "phospho jaw" did not satisfy Knef, so he sought the chemical formula for Undark from U.S. Radium. The request was denied. In 1924, he teamed with a radiation expert, who carried out an examination of the portion of Amelia's jaw that Knef had saved. Their examination revealed that Amelia had died not of syphilis but of radium necrosis. A request was granted to exhume Amelia's body in October of 1927, now five years underground. Knef's diagnosis was confirmed.

Additionally unknown to Raymond Berry as he continued to wrangle with the courts, a reporter for *The Newark Star Eagle* discovered that U.S. Radium had come to an out-of-court settlement with three families of U.S. Radium employees who had already died. The families received a total of $13,000 in 1926. These two events would eventually assist Berry in his case, but the route would be circuitous both inside and outside the courtroom.

On January 11, 1928, Grace, Edna, Katherine, Quinta, and Albina finally had their day in court when the first hearing took place.[13] Only three of them could attend the court hearing, though, as the two sisters, Quinta and Albina, were unable to move from their beds. Grace came to the court wearing a back brace that enabled her to sit in an upright position. She had now lost all of her teeth. None of the women could raise their hands to take the oath.

Grace testified that her health had been fine until she began working at U.S. Radium. Edna and Katherine concurred that they had always been healthy before working there but now could not sleep at night because of the pain. Nor could they dress, bathe, and eat without aid. U.S. Radium countered that many of the women they had hired to paint dials were "unfit" and would not have been able to take other more strenuous jobs. U.S. Radium declared that the dial painters were wrong in blaming Undark for their travails and that these women were only "declining normally" because of their already frail natural constitutions. U.S. Radium also pointed to evidence that the women were afflicted with maladies of the psyche often attributed to women at that time: hysteria, weakness of will, and general lack of ability to confront "reality." "Radium, because of the mystery that surrounds much of its actions, is a topic which stimulates the imagination, and to our mind, it is to this and not the actual fact that many of the reports of the luminous paint's effects in our paint may be attributed," declared a U.S. Radium spokesman to the *New York Times*. The Radium Girls became one of the first instances of the atomic industry mindset that would pervade industry legal defenses by minimizing or even denying outright radiation dangers in the years to come. The January hearing ended with a tentative date set for a second hearing in April.

Berry was concerned that this would be too long. The women were deteriorating quickly, and he feared that several of them might not even survive long enough to receive any benefits whatsoever. He also feared, however, that their emotional condition could be made worse by reports in the press about the proceedings. Even though newspaper reports of the plight of the five women were drawing public attention and outrage as to their treatment, Berry and some doctors worried that the news in the general media could cause undue harm to the emotional well-being of the women. In a letter to Berry, one doctor asked, "Can you get the [newspapers] to agree to keep the women out of the paper henceforth?"[14] Another doctor stated that, "I would certainly not like to have anything the matter with me and be told every few weeks that I was going to die."[15] Berry replied that he had "endeavored to discourage publicity" to protect the women.[16] The story, however, was too compelling and important to keep under wraps.

Walter Lippmann and Alice Hamilton had been in communication about the Radium Girls for more than a year by the time it came to the first hearing by Chancery in 1928. In July of 1927, Hamilton had written to Lippmann asking if his paper, *The New York World*, could be of any assistance in the matter. Lippmann had already written in defense of workers for Standard Oil who had been poisoned by tetraethyl lead a few years before, and Hamilton wondered if the same might help the dial painters' case.

In turn, Lippmann wrote to Raymond Berry asking if he could be made privy to some of the documents Berry had in his possession. At first, Berry resisted, until U.S. Radium began more stalling tactics. U.S. Radium asked to have the second hearing postponed, but one was held on April 25, 1928. The second hearing produced no settlement, and the judge set a new date for September, despite Berry's argument to the judge that the women might not live that long. Berry went so far as to negotiate with other lawyers who were willing to move their early summer cases to September to allow room for the Radium Girls in June. But U.S. Radium made the case that their witnesses would be unavailable until September anyway because some of them were going to be vacationing in Europe over the summer. The judge's decision: a hearing was scheduled for September.

Incensed by the delays of the case, Lippmann wrote an editorial on May 10 that called the proceedings and delays "a damnable travesty of justice. The women are dying. If ever a case called for prompt adjudication, it is the case of five crippled women who are fighting for a few miserable dollars to ease their last days on earth."[17] U.S. Radium countered Lippmann by having Frederick Flynn—the "doctor" who had proclaimed Grace fit as a fiddle three years earlier—hold a press conference in which he claimed that the women would indeed survive and there was *no radioactivity* shown in the tests he had run on them. Lippmann speedily responded in several more articles. Most pointedly, on May 19, he wrote, "This is a heartless proceeding. It is unmanly, unjust and cruel. This is a case that calls not for a finespun litigation but for simple, quick, direct justice."[18]

Marie Curie even responded to the plight of the women in May. She had read of the case in French newspapers and was distraught over the reports. "I would be only too happy to give any aid that I could.

[However], there is absolutely no means of destroying the substance once it enters the human body."[19] Curie knew the disease very well; she would die of it herself in 1934.

Berry, backed by the many press releases that were now garnering public outrage, managed to get a new trial date set for June. In early June, a federal judge, William Clark, agreed to mediate the case. In part, driven by the newly discovered report from 1926 demonstrating that U.S. Radium had already settled one case out of court, Berry sought a similar settlement for the five women. In the end, a settlement was reached. Each woman received $10,000 and a $600-per-year annuity "for life." In addition, husbands of the married women received a small sum "for loss of services" by their wives.

One note: Judge Clark was a stockholder in U.S. Radium.

Radium dial painting at large facilities in the United States continued until about 1940 when the companies closed their doors, but some dial painting in smaller operations continued into the 1960s. In the 1980s, several of these sites, notably the one in New Jersey but also in Ottawa, Illinois, and Westport, Connecticut, were declared Superfund sites. Subsequently, they were cleaned up by destroying the buildings, removing the old materials and soil, and trucking the radioactive debris to sites in Utah, Nevada, and other remote western regions. The cost of these cleanups ran into the tens of millions, if not more.

Grace, Edna, Katherine, Quinta, and Albina all died by 1935. Their deaths and their courage confronting the agents of their death would serve to foster significant[20] changes in industrial health law and worker compensation law: an incredibly high price to pay, a price paid by innocent women in the service of making pretty things. In many of the press reports about these women, their diseases and deaths were often blamed upon the element that caused their demise: radium. Even though many changes to policies and laws would be inspired by their horrible experiences, the people behind the radium painting industry were often as not left off the hook. Radium was declared to be the culprit, not the humans who developed it.

NOTES

1. C. Clark, *Radium Girls*, 16. Clark's text is a comprehensive history of the dial painters' cases.
2. Ibid., 15.

3. A. Bellows, Undark, http://www.damninteresting.com/?s=radium +girls.

4. U.S. Radium Corporation, http://libraries.umdnm.edu/History_of _Medicine/USRadium Corp.

5. M. Neuzil and B. Kovarik, *The Radium Girls*, 37.

6. Clark, *Radium Girls*, 17.

7. S. Hillgartner, *Nukespeak*, Chapter 1.

8. ERC student handbook, Harvard School of Public Health, www .hsph.harvard.edu.

9. Neuzil and Kovarik, *The Radium Girls*, 38.

10. Ibid., *The Radium Girls*, 37–38.

11. Ibid., *The Radium Girls*, 33–50.

12. *Time* in Partnership with CNN. "Medicine: Poison Paintbrush." *Time*, June 4, 1928.

13. Raymond Berry microfilm collection.

14. Neuzil and Kovarik, *The Radium Girls*, 44.

15. Ibid., 44.

16. Ibid., 44.

17. Ibid., 46.

18. Ibid., 46.

19. Ibid., 37.

20. Clark, *Radium Girls*, 203–210.

CHAPTER 2

On a Need-to-Know Basis: From "The Bomb" to the Uranium Frenzy to the Living Room

On August 2, 1939, Albert Einstein signed a letter that would change the world. With two other physicists, he wrote a 500-word letter to Franklin Roosevelt. Roosevelt's response would mobilize activities affecting virtually every person on earth and set into motion a changing of human minds, from the most powerful political, scientific, and military leaders across the globe to the people on the streets, in the fields, and on the oceans.

Einstein's two collaborators, Enrico Fermi and Leo Szilard, had grown very concerned over the war in Europe and wanted to convince the president to develop an atomic bomb. They enlisted Einstein because of his prominent status in such matters and his shared concern over the growing war in Europe. Further, they persuaded Alexander Sachs, an investment banker who was close to the president, to help them hand deliver the letter. With all four men in attendance, Sachs read the letter aloud to the president:

In plain and calm language, the message of the letter was clear: We believe we know how to make a bomb of enormous power, we must do so, and money will be needed, along with government support at many levels. Upon hearing the text of the letter, Roosevelt responded, "Alex, what you are after is to see that the Nazis don't blow us up." He then called his secretary, Pa Watson, into his office and said, "Pa, look into this. It requires action."[1]

Roosevelt's simple command activated what would become one of the largest and most costly efforts in American history, the Manhattan Project. From its humble beginning through the Uranium Committee (a subcommittee of the National Defense Research Committee) that

Albert Einstein
Old Grove Road
Peconic, Long Island
August 2nd, 1939

F.D. Roosevelt
President of the United States
White House
Washington, D.C.

Sir:

Some recent work by E. Fermi and L. Szilard, which has been communicated to me in manuscript, leads me to expect that the element uranium may be turned into a new and important source of energy in the immediate future. Certain aspects of the situation which has arisen seem to call for watchfulness and if necessary, quick action on the part of the Administration. I believe therefore that it is my duty to bring to your attention the following facts and recommendations.

In the course of the last four months it has been made probable through the work of Joliot in France as well as Fermi and Szilard in America—that it may be possible to set up a nuclear chain reaction in a large mass of uranium, by which vast amounts of power and large quantities of new radium-like elements would be generated. Now it appears almost certain that this could be achieved in the immediate future.

This new phenomenon would also lead to the construction of bombs, and it is conceivable—though much less certain—that extremely powerful bombs of this type may thus be constructed. A single bomb of this type, carried by boat and exploded in a port, might very well destroy the whole port together with some of the surrounding territory. However, such bombs might very well prove too heavy for transportation by air.

The United States has only very poor ores of uranium in moderate quantities. There is some good ore in Canada and former Czechoslovakia, while the most important source of uranium is in the Belgian Congo.

In view of this situation you may think it desirable to have some permanent contact maintained between the Administration and the group of physicists working on chain reactions in America.

One possible way of achieving this might be for you to entrust the task with a person who has your confidence and who could perhaps serve in an unofficial capacity. His task might comprise the following:

a) to approach Government Departments, keep them informed of the further development, and put forward recommendations for Government action, giving particular attention to the problem of securing a supply of uranium ore for the United States.

b) to speed up the experimental work, which is at present being carried on within the limits of the budgets of University laboratories, by providing funds, if such funds be required, through his contacts with private persons who are willing to make contributions for this cause, and perhaps also by obtaining co-operation of industrial laboratories which have necessary equipment.

I understand that Germany has actually stopped the sale of uranium from the Czechoslovakian mines which she has taken over. That she should have taken such early action might perhaps be understood on the ground that the son of the German Under-Secretary of State, von Weizsacker, is attached to the Kaiser-Wilhelm Institute in Berlin, where some of the American work on uranium is now being repeated.

Yours very truly,
Albert Einstein

Source: Smyth, Henry DeWolf. *A general account of the development of methods of using atomic energy for military purposes under the auspices of the United States government, 1940–1945.* Washington, DC: U.S. Government Printing Office, 1954.

had an unusually meager startup budget of $6,000, the drive to defeat the Axis powers gained a stunning momentum. By 1942, appropriations were made available to begin the intensive scientific research effort and the massive infrastructure of the Manhattan Project that eventually cost more than $2 billion by the war's end. As Corbin Allardice and Edward R. Trapnell describe in their history of the Atomic Energy Commission:

Soon, hundreds, then thousands, then tens of thousands of scientists, engineers, technicians, and craftsmen were looking and working. It was a desperate search. They began to spend hundreds of thousands, then hundreds of millions of dollars to find what they were looking for before Hitler's scientists found it first ... FDR's order led to a national investment of $2 billion by the end of 1946 and twenty times that amount twenty years later.[2]

The developing atomic mindset was not only concerned with the speed and resources provided for this endeavor, but also that it was done with such great secrecy. A dead silence was maintained even as the Manhattan Project grew to its enormous scale. By 1943, the Oak Ridge, Tennessee, processing facility, containing what was then the largest building in the world, was fully operational; the Hanford, Washington complex, employing as many people as the nation's automobile industry, was also running; and the Los Alamos Laboratory in New Mexico was assembling one of the greatest concentrations of scientific genius ever to meet in one physical space.

But very few outside these facilities knew what was going on inside, even at the highest levels of the government. Secrecy would become one of the great hallmarks of the whole atomic and nuclear development enterprise, and the fact that it remained a secret in this early stage of development is, in retrospect, mind boggling.

The idea that such a widespread effort could maintain secrecy over a period of many years is a key marker of how the atomic mindset came to be. Entire geographic and intellectual communities were willing to work in secret partnership because they believed that the effort was in support of democracy and the war effort being waged on two fronts. This is, in the context of that moment, understandable. The country had been through a deep economic depression, and jobs were now available. The war effort involved nearly every facet of American industry, and the developments in the atomic industry were part and parcel of the greater war effort. The mindset of secrecy was strongly reinforced through patriotic adrenaline and the guarantee of jobs. Questions were not welcome, nor were they likely to occur.

Even at the highest levels of government, only spotty attempts were made to find out what was really going on. When Harry Truman was still a senator in 1944 and a member of the special Committee Investigating the National Defense Program, he sent a message to the secretary of War, Henry Stimson, asking that a general go to Hanford, Washington, to investigate the project that was draining government resources at increasing rates. He was rebuffed by the war department and, by some accounts, accepted assurances that all was well with the expensive and multifaceted endeavor that was in the country's best interests. Truman backed down, as did several other congressional leaders.[3]

It wasn't until April 12, 1945, the day Franklin Roosevelt died of a massive stroke, that the now-President Truman learned of the massive Manhattan Project and the intention to develop an atomic bomb. Just six years after Einstein's letter, Truman would make the final decision to drop the bombs on Hiroshima and Nagasaki.

THE CONUNDRUM OF SECRECY: OPENNESS, OF A SORT

Almost as quickly as Einstein's letter had propelled action on building an atomic bomb, the secrecy of the bomb and some of its associated industries became known to the public through the bombs dropped on Japan. In an instant, the atomic bomb went from the biggest, best-kept secret in U.S. military history to the subject of intense public fascination, of countless articles, essays, and news reports. The potential for peaceful uses of atomic power—the idea that atoms could power cars, electric plants, and submarines—absorbed the attention of creative writers, filmmakers, and artists as well as engineers and scientists, although no scientific methods for such uses had been developed by the war's end. In 1946, President Truman established the Atomic Energy Commission (AEC), the civilian successor to the Manhattan Project, to pursue their development.

Consequently, the intense secrecy of the bomb-making effort had now become a conundrum for the government and the industrialists who wanted to pursue further development of atomic weaponry and other possible peaceful purposes, a conundrum of maintaining intense secrecy and, at the same time, letting just enough of the genie out of the bottle to engage the public's mind. On the one hand, secrecy about how to make such weapons was still of paramount concern, as the Atomic Energy Act of 1946 makes explicit with regard to fissionable production processes and materials:

Whoever, with intent to injure the United States or with intent to secure advantage to any foreign nation, acquires or attempts or conspires to acquire any document, writing, sketch, photograph, plan, model, instrument, appliance, note or information

involving or incorporating restricted data shall, upon conviction
thereof, be punished by death or imprisonment for life ...[4]

Indeed, even as late as the 1970s, the fear that someone might divulge
the "secret" of the A-bomb held sway in the government mindset, as
demonstrated in the example of Howard Morland. In 1979, using little
more than the *Encyclopedia Americana* and some other public sources,
Morland constructed a model of an H-bomb warhead out of plastic,
wood, and other mundane materials. He subsequently wrote an article,
"The H-Bomb Secret (To Know How is to Ask Why)" for *The
Progressive*.[5] The federal government blocked publication of the article,
but after a six-month legal battle, Morland was finally able to see the
article in print.

On the other hand, for civilian atomic energy to grow, the
government needed the public's assistance. The first atomic bombs
had all but depleted the country's stockpile of enriched uranium. It
takes enormous amounts of raw uranium to make the fissionable
materials necessary for bombs the size of the ones used in the war,
and it would take even greater amounts to develop the new, more
powerful bombs being planned. There were few known places in the
world where significant quantities of uranium could be mined. The
United States had drawn nearly all of its World War II uranium from
the Belgian Congo (now the Democratic Republic of the Congo),
which was still a fruitful source for the richest uranium ore (known
as pitchblende), and the Eastern European mines that Hitler had captured were temporarily available; but even with these reserves, more
uranium was needed.

To this end, the federal government openly encouraged freelance
uranium exploration. The government determined it now needed public support to find new sources of uranium. Thus, a public initiative
through atomic education and marketing was developed that, in turn,
required that the government unveil some, albeit few, of its atomic
secrets. Consequently, the federal government pursued two ways of
encouraging open support for uranium exploration and atomic energy
research and development. The first way was a call for freelance public
uranium prospecting that would take the country by storm in the 1950s.
The second way, less direct in some aspects than the overt pursuit of
uranium prospecting, was to develop a mindset through formal and
informal education of the nation's youth that would prepare future scientists and ultimately make the "atomic dream" a reality. The dream
opens with the uranium prospecting rush.

THE AMERICAN URANIUM FRENZY

In the popular mindset, a rush usually focuses on stories about pro-specting for and mining precious metals and gems like gold, silver, copper, and diamonds. Some of these stories also tell of rushes for less glamorous elements such as iron, selenium, vanadium, and other met-als necessary for industrial development. Uranium's story is of this second type and was mostly targeted to residents of the Western United States. However, the uranium rush of the mid-twentieth cen-tury was actually quite the national phenomenon, a phenomenon largely invisible to the average U.S. citizen. The story that follows is about what constituted this rush and the forgetfulness that is left in its wake.

In 1948, the Atomic Energy Commission (AEC) established a guar-anteed price for uranium, a price not to decrease for ten years. In addi-tion, the AEC began offering bonuses of $10,000 (and raised it to $35,000 for a time in the early 1950s) to individuals who found signifi-cant deposits of uranium on public or private lands. To encourage average citizens to become involved in uranium exploration, the AEC even published a small pocket-sized manual titled *Prospecting for Uranium* that cost thirty cents and sold more than 16,000 copies within a year of its 1948 release. As the booklet explains, prospectors could stake claims on private land with no restrictions and on public land if they merely paid one dollar for the claim and then took their uranium to be assayed for its radioactive quality in offices run by the AEC. Further, public lands were expanded in some cases to include some national parks and monuments, such as Mt. McKinley National Park, Glacier Bay National Monument, Organ Pipe Cactus National Monument, and Death Valley National Monument. The result of this call for uranium prospecting drew thousands of hopeful prospectors, both amateurs and professionals.

Much of the uranium frenzy took place in the American West and predominantly on the vast Colorado Plateau. There were prospectors who certainly struck it rich, and their stories are often told in popular literature and histories. The most legendary of these atomic-age prospectors was Charles Steen. Steen quit his job as an oil worker in Oklahoma and Texas in 1948 and took his wife and two sons, with a small travel trailer, into the back parts of the plateau. He searched for more than three years to find the pot of uranium ore at the end of the proverbial rainbow. After nearly running out of money to the

Drawing of a prospector's uranium mine in the American West, circa 1955.
(From *Prospecting for Atomic Minerals* by Alvin Knoerr and George P. Lutjen.
New York: McGraw-Hill, 1955.)

point that his family was literally living on dried beans, Steen discovered the magical pot—and it was a pot not of the ordinary grade of
uranium known as yellow cake, but of the much richer grade, pitchblende, that contained upward of 70 percent pure uranium. Steen
became fabulously wealthy and by one account became the richest
man in Utah, garnering a fortune of about $130 million by 1955.
Steen would go on to squander his riches over the next decade
through lavish living and poor investments, but his story sounded
the siren call that money was indeed to be made in uranium.

There were several other mother lode stories like Steen's in which
prospectors made millions and others reaped profits in the tens
of thousands. Most prospectors, however, were not so lucky. Nevertheless, their efforts fueled the rush mindset that traveled far beyond
the Colorado Plateau. Some most interesting pieces of evidence are
found in the *New Yorker* magazine "Talk of the Town" articles published between 1948 and 1953—right in the heart of the rush frenzy.

In the August 27, 1949, *New Yorker*, Brendan Gill tells of being
intrigued by a sale at Abercrombie & Fitch for Sniffer Geiger Counters that were "for the benefit of sportsmen who want to take up uranium prospecting as a sideline."[6] The Geiger counters so interested
Gill that he investigated more fully the extent of uranium prospecting
fever in Manhattan and discovered that prospectors were not only

going to Colorado in search of uranium, but they "are currently clambering up schist and down shale within a hundred miles from here."[7] He subsequently visited the local AEC Office in the Justice Department on Columbus Avenue and there encountered a Mrs. Muriel Mathez, who was the chief mineralogist of the Raw Materials Operations Laboratory of the AEC. She said that the uranium samples that they received on a daily basis from regional uranium hounds were mushrooming so quickly that they had received 2,500 samples of rock since the uranium-hunting program began since 1948. "We're swamped here as are the offices in Washington," Mrs. Mathez said. In a moment of some lament, though, she added that "Ninety percent of the samples never get past our first Geiger counter test for the simple reason that they're not radioactive. Heaven knows why people send them to us."[8]

Readers of the *New Yorker* continued to find uranium fever intriguing well into the 1950s, and sales of Geiger counters and other tools of the trade remained steady. In a January 19, 1953, "Talk of the Town," story Rex Lardner wrote that a company in Manhattan called the Radiac Company—a wholesale supply company—had such a demand for Geiger counters that the company had decided to go into the retail business. The sales manager at Radiac, James Mitchell, described the intense demand for prospecting equipment but then described another use of the Geiger counters, a use that will come up again later in regard to promoting the atomic mindset to youths. Mitchell pulled a vase from a shelf and proceeded to flip a switch on one Geiger counter, a Classmaster. Immediately, the machine began to click. Mitchell explained that the Classmaster was a potential visual training aid that educators could use to teach about radioactivity— what he called the "Fourth R" of literacy (added to readin', ritin', and 'rithmetic). He further explained that the vase was radioactive due to the uranium salt paints on the decorative exterior. "Pottery isn't being painted with uranium salt anymore," he said, "because the AEC won't allow it." He continued, "The hands on some people's watches are radioactive ... glassmakers in Europe used uranium salts to give color to stained-glass. The windows of the Cathedral of St. John the Divine are radioactive. But harmless."[9]

Harmless is hardly the right word. While raw uranium is not nearly as powerfully radioactive as, say, radium, which killed the Radium Girls, it can still pose a health risk, especially if it is of a high-grade ore when the dust or minute particles are inhaled or ingested. However, the danger from exposure to radioactive substances was downplayed during the uranium rush. This was partly because the full extent of that danger was yet unknown. In addition, the embargo on communicating atomic information legislated through the Atomic Energy Act of 1946 had the effect of hushing the discussion of what was already known. The government's interest in acquiring uranium and its dangers simply paled before what they hoped to gain through increased uranium exploration.

FUN, GAMES, AND LEARNING ABOUT URANIUM PROSPECTING

Meanwhile, the same spirit of innocence and optimism that animated the Radiac sales manager helped bring the fourth R into the realm of games, toys, and science kits. Board games like the highly popular Careers, first introduced in 1955 by Parker Brothers, had players compare eight career paths in terms of the fame, fortune, and happiness they could provide. One of the careers in the original version was uranium prospecting. Uranium Rush, an award-winning game of the same era, was modeled after the uranium bonuses and government-prescribed procedures for staking a claim. Players spun a dial that designated where to go on the board to explore for possible uranium; once you landed on one of the geographic points represented on the board, a battery-run "Geiger counter" was placed on the point of the claim and, if it lit up and buzzed, you won $50,000 from the government bank. The winner of the game was the prospector with the most money—"just as in real life," as the promotional materials for the game proclaimed.

Touted as fun and educational, these and other board games were joined by the even-more-educational uranium-prospecting science kits. These included actual Geiger counters, cloud chambers for running experiments, and samples of uranium and radium to be used by the budding young scientists. Possibly the most elaborate of these kits was the Gilbert U-238 Atomic Energy Lab circa 1950–51. It was an expensive toy, costing about $50, and included a Geiger counter, cloud chamber, electroscope, and four types of uranium ore. In addition, this kit had three other significant pieces: (1) the AEC's *Prospecting for Uranium* handbook, (2) a rather sophisticated manual that included instructions for conducting experiments with the contents of the kit and readings on the nature of atomic physics, and (3) a reprint of a 1947 full-color comic book titled *Dagwood Splits the Atom*.

Prefaced by a letter from General Leslie Groves of the Manhattan Project, the 37-page comic employs Mandrake the Magician as primary narrator and includes secondary characters such as Blondie, Popeye, Wimpy, and others to explain, to Dagwood and to kids, the history and nature of atomic fission. Straightforward instructions, the inclusion in the kit of actual pieces of uranium ore of different grades, and the popular appeal of comic characters were a serious attempt to employ humor and nonthreatening antics to spur youths into action. The comic book concludes with a textbook-like set of questions and answers, and it overviews the potential for future scientists to develop miraculous, peaceful products for industry, medicine, and agriculture through the atom. The kit had a clear pedagogical intent: to demonstrate that if Dagwood can split the atom, then anyone can. Any aspiring young scientists would feel encouraged to make the dream come true. As Wimpy exclaims near the end of the comic, "Furthermore, Popeye, don't forget scientists must learn more and more about the atom before they can harness it to make it actually go to work, BUT THEY WILL!"[10]

Not only was *Dagwood Splits the Atom* included in the U-238 Atomic Lab, but, as Paul Boyer recounts in his book *By the Bomb's Early Light*, it was also handed out free at expositions promoting the peaceful uses of atomic energy such as the Man and the Atom Exhibit of 1948 in New York's Central Park sponsored by General Electric and Westinghouse. "Over 250,000 copies were distributed, leading GE to order a further printing of several million."[11]

Within ten years of Dagwood's amazing feat, the great uranium rush was nearly over as uranium stockpiles increased, new ore deposits were discovered around the world, and the price plummeted. Boom had again, as in almost all mining industries, finally become bust in a very short period of time. Nevertheless, the atomic mindset had been well established, and it was energetically trying to turn from a mindset of horror and destruction to a mindset of peace.

NOTES

1. T. Zoellner, *Uranium*, 38.
2. C. Allardice and E. Trapnell, *The Atomic Energy Commission*, 8.
3. Ibid., 21–23.
4. 79th Congress, G. Udell (compiler), *Atomic Energy Act of 1946*, 13.
5. R. Del Tredici, *At Work*, 129–131.

6. B. Gill, "Search," *New Yorker*, 1949, 19.
7. Ibid.
8. Ibid.
9. R. Lardner, "The Fourth R," *New Yorker*, 1952, 19.
10. King Features Syndicate, *Dagwood*, 27.
11. P. Boyer, *By The Bomb's*, 296.

CHAPTER 3

"Only Answered in the Stars": The Human Testing of the Bikini Islanders

"Toward the Bikini Climax," read the headline of a pictorial essay in the Sunday, May 26, 1946, *New York Times*. "From Bikini to Washington, in ports from the Atlantic to Pearl Harbor, in factories throughout the nation, at laboratories and flying fields from Tennessee to New Mexico, preparations for the greatest single experiment in all history have reached the final stage," the article proclaimed.

> When an atom bomb drops upon the guinea pig fleet deployed in Bikini Lagoon next July, its effects will be more exhaustively gauged and measured, photographed and reported, than any previous event, either natural or man-made. Here is a tentative line-up of what will be at Bikini: 42,000 personnel, 100 target ships, 12 small target craft, 123 operational ships, 139 airplanes, 7,000 instruments to photograph the event and record the radioactivity, not including 12,000 radioactivity indicating security badges, and 4,800 animals, including 4,000 white rats, 200 goats, 200 pigs, and 400 other small animals. The immediate purpose of all these men and all this equipment is to answer the broad question: Does the atom bomb make conventional navies obsolete? Yet even broader scientific questions, concerning temperatures, pressure, and speed of radiant forces normally existing only answered in the stars will be answered at Bikini.[1]

Eight captioned photographs depicting the crews, equipment, and targets of the massive "experiment" accompanied the short article. One picture shows Major Thomas W. Ferebee and Captain Kermit K. Beahan, described as "atomic bombardiers of Hiroshima and

Nagasaki," plotting a course as they trained for the tests. Another photograph, unfortunately in black and white, shows "Guinea Pig No. 1—The U.S.S. *Nevada* that will have the place of honor in the bulls-eye shown in the picture at the top, right, and presumably be blown to bits. She has been painted orange-red." And a third photo shows the radioactivity-indicating badges worn by the military personnel. The badges depict a mushroom cloud and crossed roads leaping like lightning bolts from the aura surrounding the cloud. Operation Crossroads, as it was called, was in full swing and ready to produce some of the largest atomic explosions ever witnessed. The government, the media, scientists, government officials, the military, and even the public waited with bated breath for the summer's "experiment."

There was no secrecy guarding this event—well, almost none. Mostly there was manipulation. The Bikini Islands—part of the Marshall Islands, located midway between Hawaii and Australia in the southern Pacific Ocean—are home to the Bikinian people who, in March 1946, were asked by U.S. Navy Commodore Ben H. Wyatt, the military governor of the Marshalls, to leave their islands, only temporarily, so that the U.S. could test atomic weapons there "for the good of mankind and to end all wars." As Jack Niedenthal explains the moment in *For the Good of Mankind: A History of the People of Bikini and Their Islands*, "King Juda, then the leader of the Bikinian people, stood up after much confused and sorrowful deliberation among his people, and announced, 'We will go believing that everything is in the hands of God.' "[2]

Whether God was involved in the destiny of the Bikinians is a matter for debate. What is certain, however, is that the U.S. government and military would have much to do with the fate of the Bikini Islanders over the next 65 years. For six decades now, the people of Bikini have been shuttled repeatedly from one atoll to another across the 357,000 square miles of the Marshall Islands chain. Their trek began in March 1946 in preparation for the two atomic bombs, dubbed Able and Baker, tested by the U.S. military during Operation Crossroads. A U.S. Navy landing craft transported the Bikinians 125 miles eastward to Rongerik Atoll. This new island "home" was much smaller than Bikini and had fewer food sources. They were given several weeks of food supplies by the Navy and then left to fend for themselves. The odyssey they had embarked on was one that the ancient Greek sailor Odysseus himself might think twice about. At least he had occasional agency and choice along the route. The Bikinians had none of this.

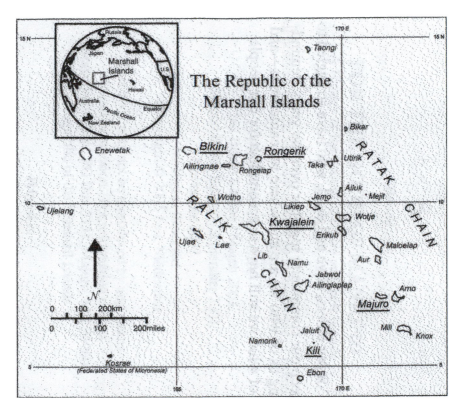

Map of Marshall Islands. (Courtesy of Sasha Davis)

Having removed the Bikinians, the Army and the Navy were anxious to make up for lost time: They had an atoll or two to blow up. The preparations for Operation Crossroads had been intense and, as shown by the *New York Times* photo spread, involved the military, the scientific community, the press, and, of course, the Bikinian people agreeing to cooperate. In fact, the fateful meeting between the people and Commodore Wyatt was reconstructed and filmed by the media for posterity and to announce to the world that the decision had been made by the Bikinian people. The indigenous population, however, did not know that other decisions regarding their fate had been made in secret—decisions made by U.S. government officials and scientists to use them as human subjects in the ongoing study of radiation and its effects upon humans and natural environments.

In his published diary *No Place to Hide*, military physician David Bradley, who monitored and studied radiation levels in the air, the water, and the bodies of personnel, offers a personal account of what

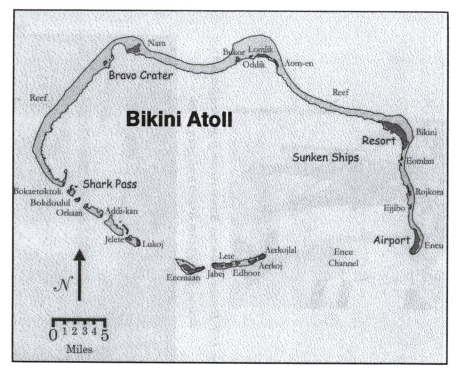

Map of Bikini Atoll. (Courtesy of Sasha Davis)

it was like on those fateful days in July 1946 as preparations were made
to study the effects of the atomic explosions.

> By 5:30 (July 1) in the first light of dawn we were airborne, climb-
> ing slowly to 8,000 feet and swinging away to the northwest.
> Dawn came and passed unnoticed . . . There was much to do . . .
> First there were the instruments on the navigation table. Turning
> them on, I tested each one with a small pocket radium source . . .
> Next were the film badges, protective goggles and gas masks to
> pass out to all hands. In spite of the distracting smell of steak
> coming out from the galley, the crew welcomed their issue of
> strange gear. The little films, worn in the pocket, were sheer mys-
> tery to them; the goggles meant total darkness during the actual
> flash; but the gas masks they understood . . . By the time we had
> cleared away the steak and French fries, washed them down with
> a cup of scald and topped it off with a cigarette, we were dropping
> down through the cottony cloud layers to Bikini. The live fleet
> was strung out in several columns under us, and some twenty

miles away the target fleet was visible bunched up at one end of the lagoon.[3]

Just moments after sighting the lagoon, the intercom in the aircraft announced that Able had been dropped. But witnesses in the aircraft, Bradley recorded, at first detected nothing—there was no visible or audible evidence of explosion.

> The ball of fire which had fused several acres of the New Mexico desert into glass and turned Hiroshima into a symbol of man's inhumanity burst over the target ships, seared the paint from their decks and melted down their masts, but at twenty miles gave us no sound or flash. Coming out from under we could discern no change in the world. Had it been a dud? . . . Then, suddenly we saw it—a huge column of clouds, dense, white, boiling up through the strato-cumulus, looking much like any other thunderhead but climbing as no storm cloud ever could. The evil mushrooming head began to blossom out. It climbed rapidly to 30,000 or 40,000 feet, growing tawny pink from oxides of nitrogen, and seemed to be reaching out in an expanding umbrella overhead.[4]

Able was dropped from a B-29 bomber and, to the dismay of the scientists and military personnel, it missed its target—the bright orange U.S.S. *Nevada*. Bradley's description of a "possible dud" was indeed somewhat accurate. Despite his description of the explosion's amazing aftermath, Able was disappointing. Much of the target fleet was still afloat, and although some significant radiation readings were recorded, the event was not as devastating as the military and scientific observers had hoped. In three more weeks, though, the Baker test would provide a second chance.

"Baker Day—Thursday, July 25, 1946," David Bradley's diary began.

> Ten seconds . . . five seconds . . . four seconds . . . three . . . two . . . one. What his [the detonator's] last word was I have no idea, nor can I tell what color the flash was. To me it was red; Lars swears that it was white . . . A gigantic flash—then it was gone . . . Then a huge hemispheric mushroom of vapor appeared like a parachute suddenly opening. It rapidly filled out in all directions until it struck the level of the first layer of clouds, about 1,800 feet.

Operation Crossroads Baker Blast, 1946. (Defense Threat Reduction Agency (DTRA))

> Here, as though striking a layer of plate glass ... this cloud of vapors spread out by leaps and bounds beneath the clouds. I remember being alarmed lest our plane be overtaken and smashed by it.[5]

Baker by all accounts was much more successful than Able. Not dropped from on high but planted under the ocean and detonated by electric remote controls, it was right on target: an enormous explosive sea-mine that rose thousands of feet into the air. While some ships went under immediately, in the next several weeks, more ships would slowly sink from the damage they had incurred. Many of the target-area ships remained afloat, however, and herein lurked a danger. The Baker and Able explosions had produced levels of radiation that made boarding the still-floating but damaged vessels risky. The Able and Baker blasts demonstrated America's atomic weaponry and helped to create the greatest radiation Petri dish of all time—here was an opportunity to measure radiation on objects, people, and animals that had

Keep Clear. Danger! Very Radio-active. USS *Catron* damaged in Operation Crossroads atomic bomb test in the Marshall Islands, 1946. (Defense Threat Reduction Agency (DTRA))

been on the ghost fleet during the blast, and the American navy men who boarded the floating derelicts with mops and buckets in hand in an effort to clean the radioactive fallout from the decks of the ships.

The more seaworthy vessels were in fact boarded, and radiation levels were measured. The Geiger counters sang like flocks of crazed songbirds, clicking raucously in a steady stream rather than with the staid tick-tick-tick of moderate radiation. Some personnel attempted to physically scrub the vessels clean of radiation. Bradley describes these efforts in his diary: "The decks had been sluiced with water. When this caused no reduction in radioactivity buckets of soap were broken out. Still the same result. The time-tested formula of a 'clean sweep-down fore and aft' is out of date in the atomic age."[6] Thus the target fleet was declared unsalvageable, and any ships that didn't list and sink on their own in the weeks that followed were towed to sea and dropped down into the depths of the Pacific. This was the case with some of the active fleet, too, as these vessels could not be scrubbed free of radiation, either.

Expected radiation levels in the natural environment also proved to be more stubborn than officials had imagined. Even two weeks after the second blast, the radiation readings at the bottom of the lagoon were high. "I think none of us expected the bottom to show such intensities at this late date [August 9]," Bradley wrote.

> It poses the interesting question of whether or not the fish which live by browsing on coral and algae may ultimately be destroyed by the radioactive materials in the their bodies. And if so, what will become of the larger fish which live off the smaller fish; will they too become infected and pass on the plague of radioactivity? That is a study which would occupy at least a year out here.[7]

In fact, scientists are still studying the effects more than half a century later.

As scientists and military brass studied the blasts' effects, enlisted men learned through observation that radiation was a menace. While on board the soon-to-be-scuttled U.S.S. *New York*, a sailor responded to a question about the effects of radiation by exclaiming, "Ahhh, it don't matter, Chief . . . I got so many of them Geigers runnin' round inside me now that you can see me all lit up at night like the Statue of Liberty." Another sailor added, "That's right. Since I been out here I've grown seventy years. I don't care if I don't never see another woman. I'm a goner. I'm hopeless."[8]

These observations offered Bradley an opportunity to explain to the men how radiation behaves in the human body. Such explanations were uncommon due to the top-secret climate of atomic testing. But Bradley, in conversations he had with the sailors he worked with on-board their base ship between visits to the afflicted vessels, felt that the men should at least know what science knew about radiation. Science still had a long way to go in learning about radiation effects, yet Bradley revealed that scientists actually knew a fair amount, even in 1946. He described how radiation works from the outside in, lodging itself in bones and tissue. Ominously reminiscent of the Radium Girls,

Bradley knew that plutonium from the bombs was "roughly the same as that of radium. It lodges in the bones, destroys the blood producing marrow, and may kill its host by wrecking his red and white blood cells; or, if he survives that early period, he may die years later from bone tumors."[9]

Wanting at least to give the men some of this knowledge in the hope that it might help them protect themselves to some extent, Bradley explained that Geiger counters were important because they could forewarn and tell where not to go or when to get out of a place. But one astute sailor asked, "And what happens when that little box gets snaffued?" Bradley responded, "That is where the film badges come in. The film is exposed just like X-ray film and we can tell how much you've been getting." To this, one of the sailor's replied, "Sure, but that only tells you afterwards. A lot of good that is when you've been fried all day in X-rays."[10] So now they knew why they wore the film badges—the same badges that workers at Alba Craft, Savannah River, Los Alamos, Oak Ridge, Hanford, and other atomic development installations were never told about. They just wore them and turned them in. No questions asked. But at least these men knew what the badges were used for, thanks to Bradley, who broke the code of secrecy with human subjects.

Back at the Rongerik atoll, the Bikinians were attempting to learn how to live on their much smaller home. Not only was their land area diminished about six-fold, but the food offered by the atoll was less abundant and of a different makeup than they were accustomed to. In addition, the housing consisted of temporary shelters and offered few amenities; and according to traditional beliefs, this atoll was inhabited by evil spirits. In sum, the new home was hardly a home at all but merely a temporary, haunted stopping point—a stopping point the Bikinians believed would not last long as they waited for the promised return to Bikini.

By March of 1947, the United Nations had declared the whole of Micronesia a Strategic Trust Territory (STT) and assigned the United States management duties. The STT agreement, among other provisions, called for the United States to "promote the economic advancement of the Marshall Islanders . . . and protect the inhabitants against loss of lands and resources."[11] In the long term, this provision for "advancement and protection" would prove to be a challenge to define. Staying in one place was not a safe option for the Bikinians, and protecting lands and resources would just not happen. In the short term, the STT had made promises, but none to address the issue of using the Bikinians and other Marshall Islanders as human test subjects. That promise, if it had been made, would have been kept under lock and top-secret key.

The Atomic Energy Commission (AEC), having been recently and contentiously established in Washington, was charged with making atomic energy development seem like an everyday activity—that it, as Eileen Welsome remarks in her *Plutonium Files*, to make "atomic energy as unthreatening as electricity."[12] But one of the activities that needed to remain undisclosed in this safe development landscape was human testing. Revealing the truth of human testing could ruin the positive tone of atomic development for everyday Americans. "[P]ublic disclosure would have damaged the commission's bomb-building program and its efforts to build a civilian nuclear power industry."[13] The dual purposes of atomic development, for war and for peace, were now ensconced and would remain linked—all the way to nuclear green.

A prime example of this linkage is seen in Dr. Joseph Hamilton, a Manhattan Project physician involved with the study of radiation on humans during the development of the Manhattan Project. Hamilton saw his opportunity to study radiation effects on humans by studying the Bikinian people as too good to be true. The Bikini population would serve his interests well for a few reasons. One was that he could bypass injecting radioactive materials into humans. With the Bikinians, he wouldn't need to inject; the levels of radiation in the environment would be high enough. In addition, Hamilton had become interested—again through the Manhattan Project—in how radiation could be developed as a type of chemical weapon.

Upon returning to the United States from Operation Crossroads, Hamilton filed a report with the Department of Defense in which he claimed that "fission products can not only be used to impair or destroy civilian and military personnel, they can be used to contaminate soils and food chains and thus impair or destroy food production systems."[14] He explained further that the best way for the United States to protect itself against other countries using radiation as a weapon was to implement a "thorough evaluation of an understanding of the full potentiality of such an agent ... such studies must be made by means of large scale experimentation, as well as laboratory research, it will be essential that there be available in some isolated

region an extensive proving ground."[15] Hamilton, with a relatively
few strokes of the pen, had played a prominent role in ushering in
the scientific atomic mindset and establishing it as a new paradigm.
Scientists, protecting national security, could now legitimately study
whole populations without their consent. In addition, field trials and
laboratory research would have to be conducted to make the research
comprehensive and scientifically valid. But most importantly, this
research would be expensive—it would require significant financial
support. Thus, Hamilton pioneered a new archetype of scientific
research: the federally funded researcher. A golden age for science
was in the making.

At the time of Hamilton's reports, the unwitting guinea pig former
residents of Bikini had more fundamental concerns than being subjects of panoptic scrutiny. In May 1947, a fire ravaged Rongerik,
destroying many of the staple food crops, especially coconuts. A team
of U.S. investigators visited the atoll and was so shocked by the conditions that they called for immediate relocation of the Bikinians. Syndi-
cated columnist Harold Ickes observed the situation, too, and
reported, "The natives are actually and literally starving to death."[16]
In January 1948, further requests for aid were made by anthropologist
Leonard Mason. He was stunned by the conditions and called for a
medical officer and food to be sent, pronto. Later that year, the
United States began medical testing to study radiation effects on
inhabitants. These tests were expanded through the auspices of the
AEC in 1951 to create the first large-scale human population study
of genetic effects of radiation. These tests would continue for decades.
Secrecy in the service of national protection kept the Bikinians igno-
rant of the purposes of such data collection.

By November 1948, the Bikinians moved to Kili Island. After painful deliberation, they chose Kili because it was not ruled by a paramount king and it was uninhabited. Unfortunately, the island had no
lagoon for fishing, was unprotected from the winds and surf of the
Pacific, and presented inhospitable conditions for outrigger canoes.
Kili had an even smaller landmass than their two previous "homes."
While the Bikinians had now lost their food supply from the ocean,
the STT gave them a 40-foot boat as compensation. The boat helped
for a while, but in 1951, it was wrecked in a storm.

During this period, the Russians developed their own atomic bombs
and, in 1949, exploded the first one. Once again, the United States was

compelled to match tit for tat, and atomic tests became more frequent in the Marshalls: Between 1946 and 1958, the United States detonated 67 atomic bombs in the water, in the air, and on land. Few, if any of these tests, however, had the impact of the bomb detonated in the early morning of March 1, 1954.

In 1952, the Russians had one-upped the Americans by detonating a bomb even more powerful than Abel or Baker—a hydrogen bomb. It was purported to have hundreds of times the power of the bombs dropped on Japan. The Americans were not to be outdone. So two years later, the Americans demonstrated once again their nuclear superiority by exploding "Bravo" just northwest of the Bikini Atoll. Bravo—a hydrogen bomb with explosive power 1,000 times greater than the atomic bombs Fat Man or Little Boy that had been dropped on Hiroshima and Nagasaki in 1945—created such a massive blast that the residents of the Rongelap Atoll, about 100 miles to the south, saw "two suns" rise that morning. And some uncommon precipitation fell from the sky: radioactive fallout in tremendous amounts that came down over hundreds of square miles of ocean and atoll as snowlike ash. A Japanese fishing boat with 23 fishermen on board witnessed the ash cover their boat an hour and a half after the blast. Many of them got nauseous and one died. The people of Rongelap suffered from nausea, skin rashes, and vomiting. It would be two days before they received any medical assistance.

Human testing was not to be derailed though, and some U.S. scientists quickly evacuated some of the islanders, moved them to other islands for further study (e.g., Ailinginae, Bikar, Rongelap, and Rongerik) that were now radioactively contaminated. No forewarning of the Bravo blast had been given, so these human subjects were perfect specimens, as they had stayed right near the bulls-eye. Even if they had been warned, it would likely have made no difference. They, too, had no place to hide, just like Bradley's men in 1946.

Due to the intensity of the blast and some unfavorable weather conditions, the fallout was so intense after Bravo that the Bikini, Rongerik, Utrik, Ulelang, and Likiep atolls were all declared off limits by late March, thus forcing inhabitants to move once again. This time they were more scattered about the different atolls and islands. The fallout, in the meantime, made its presence known in 1955 as far away as Chicago when radioactive rain was detected there.

Twenty years later, Nelson Anjain, a resident of Ronjelap on that fateful day in 1954, wrote a letter to one of the doctors, Robert

Conrad, who had been commissioned by the AEC to test the inhabitants of the island for radiation poisoning following the Bravo blast. Nelson's memory of that day, and of the continuous testing of the Bikinian people, came with words, unminced:

Dear Dr. Conrad,

I'm sorry I was not at home when you visited my island. Instead, I have spent the past few months traveling to Japan and Fiji learning about the treatment of atomic bomb victims and about attempts to end the nuclear threat in the pacific.

Since leaving Rongelap on the peace ship Fri, I have learned a great deal and am writing to you to clarify some of my feelings regarding your continued use of us as research subjects.

I realize now that your entire career is based on our illness. We are far more valuable to you than you are to us. You have never really cared about us as a people—only as a group of guinea pigs for your government's bomb research effort. For me and the people of Rongelap, it is life which matters most. For you it is facts and figures. There is no question about your technical competence, but we often wonder about your humanity. We don't need you and your technical machinery. We want our life and health. We want to be free. . . .

As a result of my trip I've made some decisions that I want you to know about. The main decision is that we do not want to see you again. We want medical care from doctors who care about us, not about collecting information for the US government's war makers.[17]

Over the next decade, the Bikinians managed to forge an agreement with the United States whereby the Bikinian people (without legal representation) turned over full-use rights to Bikini in exchange for $25,000 cash and a $300,000 trust fund that provides about $15 per year to each resident through interest accrual. In addition, the Bikinians were given full-use rights to Kili. Life on the small island continued to be difficult, however, as the rough seas around the unprotected island, combined with a growing population, made food subsidies like rice and canned goods from the USDA a necessity.

In 1967, scientists inspecting Bikini made a surprising announcement: The well waters on Bikini were now safe! President Johnson used the occasion to announce in the *New York Times* that the

Bikinians could now return to their beloved homeland and build "a new and model community . . . with all possible dispatch."[18] Bureaucratically speaking, "all possible dispatch" is a relative term. When action needs to be taken, it must find its way from the White House through the Congress, across to the military, and through an exit to an entity outside the country's borders to negotiate with a territorial board.

Understood within the greater scheme of things, this process actually moved along with at least some dispatch. New coconut trees and some buildings were planted over the next few years, and in 1972, three extended Bikinian families and 50 Marshallese workers moved back to Bikini. More islanders continued to resettle on Bikini until 1975 when, not surprisingly, scientists monitoring radiation levels discovered that higher levels existed on the island than originally thought. The Department of the Interior declared, "Bikini appears to be hotter or questionable as to safety."[19] Some wells were found to be unsafe, as were some locally grown foods and native crabs. One researcher argued that the readings were "probably not radiologically significant,"[20] but by this point the Bikinians were wary of such qualifications. In October 1975, the Bikinians filed a lawsuit against the United States demanding a complete scientific survey of the island and the northern Marshalls. It would take three years of red tape before the survey would begin. The Bikinians would remain on the island during that time, awaiting the survey's results.

During the red tape hiatus, some independent and Department of Energy (DOE) scientists actually managed to conduct some testing. Finding unacceptable levels of radioactivity, the DOE advised only "one coconut per day" as an acceptable ration. Then, in April 1978, U.S. physicians found radiation levels of the now 139 residents to be beyond acceptable levels. Officials of the Interior Department called the levels "incredible" and decided to move the residents off the island in 90 days. Move them they did, once again. Ironically, once they were removed, the full survey won in the lawsuit finally commenced: too little, too late.

SELF-GOVERNMENT, MORE LEGAL BATTLES, AND COMMEMORATION

In 1978, the Marshall Islands Constitutional Convention—a governmental result, in part, of the establishment of the Congress of Micronesia in 1965—adopted the nation's first constitution. The Constitution of the Marshall Islands is indeed reflective of many constitutions of democratic nations, especially in terms of its structure. It establishes guidelines and rules for juridical, legislative, and individual rights and goes into great detail about the makeup of governmental agencies and regional governmental authorities. Interestingly, it also directly comments upon some issues not always as clearly present in, for instance, the U.S. Constitution.

Capital punishment, for instance, is banned: "No crime under the law of the Republic of the Marshall Islands may be punished by death" (Art. 1, Sec. 6.1). Conscription into the armed services is also explicitly limited: "No person shall be conscripted to serve in the armed forces of the Republic of the Marshall Islands except in time of war or imminent danger of war as certified by the Cabinet" (Art. 1, Sec. 8.1). And there is a lengthy section devoted to "Just Compensation" that provides nine guidelines for compensation in regard to land rights. All taking of land is considered unlawful if not accompanied by "an order by the high Court providing for prompt and just compensation" (Art. 1, Sec. 5.4), and "Whenever the taking of land rights forces those who are disposed to live in circumstances reasonably requiring a higher level of support, that fact shall be considered in assessing whether the compensation provided is just" (Art. 1, Sec. 5.6).[21] Taking and compensating: two issues all too familiar to the people of the Marshall Islands.

Most telling of the spirit of the Marshallese and Bikinian peoples, however, is inscribed in the Constitution's Preamble. In five short paragraphs, the people of the Republic of the Marshall Islands offer themselves to the rest of humanity through a voice of strength, openness, and goodwill to all:

CONSTITUTION OF THE REPUBLIC OF THE MARSHALL ISLANDS

Preamble

We, the People of the Republic of the Marshall Islands, trusting in God, the Giver of our life, liberty, identity and our inherent rights, do hereby exercise these rights and establish for ourselves and generations to come this Constitution, setting forth the legitimate legal framework for the governance of the Republic.

We have reason to be proud of our forefathers who boldly ventured across the unknown waters of the vast Pacific Ocean many centuries ago, ably responding to the constant challenges of maintaining a bare existence on these tiny islands, in their noble quest to build their own distinctive society.

This society has survived, and has withstood the test of time, the impact of other cultures, the devastation of war, and the high price paid for the purposes of international peace and security. All we have and are today as a people, we have received as a sacred heritage which we pledge ourselves to safeguard and maintain, valuing nothing more dearly than our rightful home on the islands within the traditional boundaries of this archipelago.

With this Constitution, we affirm our desire and right to live in peace and harmony, subscribing to the principles of democracy, sharing the aspirations of all other peoples for a free and peaceful world, and striving to do all we can to assist in achieving this goal.

We extend to other peoples what we profoundly seek from them: peace, friendship, mutual understanding, and respect for our individual idealism and our common humanity.

Given the establishment of a constitution and a well-defined governmental structure, the Bikinian people moved forward in the pursuit of compensation and recognition by the world of their history and what they needed to pursue the goals of living in peace and harmony among themselves and with the rest of humankind. Thus, the next three decades were full of continued legal battles to receive the just compensation for their populace to prosper and grow. One suit filed against the United States in 1981 lingered in the court system for six years and was finally dismissed in 1987. They did receive two trust funds to compensate for the lands they had surrendered to nuclear testing, but these were paltry compared to the needs of education, food, and other essential elements of a strong economy and way of life, modest as their requests were. The lawsuits continue to this day and have even gone to the U.S. Supreme Court—alas, to no avail. The court dismissed the case in July 2010, but more work continues in their odyssey toward just compensation for the beloved lands they lost to the atomic mindset.

Nevertheless, the trust funds that were awarded allowed for some new initiatives to begin and, under the stewardship and advocacy of a permanent trustee—Jack Niedenthal—the Bikinians have managed to create successful schools and educational opportunities for their children, both in the islands and abroad. (See the accompanying interview with Niedenthal for more specific elaboration of these activities over the past 25 years).

Further, the Republic and the local government of Bikini submitted an application to UNESCO requesting that Bikini become a World Heritage Site. This distinction promises to show to the world the tremendous human sacrifices of their history but also provides opportunities to bring in badly needed resources through such things as archaeological research of the naval artifacts below the atoll's waters and a burgeoning scuba visitation site for adventurous divers (Table 3.1). In the opening of the application's 83 pages of text and photographs, Senator Tomaki Juda eloquently makes the case for the World Heritage Site distinction:

> We, the people of Bikini, will always remember Bikini Atoll as our beloved homeland and will always feel pain for what we have lost. As a World Heritage site, Bikini Atoll will remind all of us, around the world, of the need for global peace and the elimination of weapons of mass destruction. Bikini Atoll may then actually fulfill the promise for which we reluctantly left our homeland, more than 64 years ago, "for the good of mankind and to end all world wars." In support of this nomination and the ongoing protection and management of Bikini Atoll, the community will move to establish the Bikini Atoll Conservation Management Board and undertake to develop the resources and partnerships required to effectively implement the Bikini Atoll Conservation Management Plan. We will make every effort to tell the story of Bikini to visitors, to people around the world, and most of all to our children – "for the good of mankind"— and may we never forget.[22]

Yes, never forget.

Table 3.1
The Bikinian Odyssey, a Chronology of Major Events of the Bikinian People, 1946–2011

Date		
March 1946	Bikinian people meet with Commodore Wyatt and agree to move from Bikini Island "for the good of mankind" 125 miles east to uninhabited Rongerik Atoll.	Food supplies are scant and the island has few food sources. Starvation begins to occur and plans are made to move to another atoll.
July 1946	The atomic bomb tests, Able and Baker, are exploded at the Bikini Atoll with great public fanfare and media attention.	The United States installs target fleet, and hundreds of square miles of the Marshall Island are destroyed and contaminated with extensive radioactive materials.
March 1947	Micronesia is named a Strategic Trust Territory (STT) by the U.N. and put under U.S. management. This is the only STT ever created by the U.N.	The agreement calls for the United States to "promote the economic advancement of the Marshall Islanders ... and protect the inhabitants against loss of lands and resources."
May 1947	A fire destroys many of the coconut trees on Rongerik, and a team of U.S. investigators determine that the Bikini people should be moved immediately.	Reporter Harold Ickes writes in a syndicated column: "The natives are actually and literally starving to death."
January 1948	Anthropologist Leonard Mason visits Rongerik and is stunned by the state of the Bikini people.	Mason requests a medical officer and food supplies to be flown immediately to Rongerik.
March 1948	Bikinians are transported to Kwajalein Atoll, where they are put into tents and given a strip of grass adjacent to a U.S. military airstrip.	The Bikinians are very unhappy with the situation and choose to be relocated to the Kili Islands.

Summer 1948	Medical tests of island inhabitants are begun to study radiation effects on humans. These continue for several decades.	In 1951, tests are expanded by the AEC to create the first human population study of genetic effects of radiation exposure. The Bikinians know nothing of these purposes until 1982.
November 1948	The Bikinians move to Kili Island—only a single island of one-third square mile of land area and no lagoon, as compared to the 23 islands and 3.4 square miles of land in the Bikini Atolls.	Lack of natural food resources causes more near starvation. The STT provides them with a 40-foot boat to transport supplies, but it is wrecked in a 1951 storm.
1947–1954	More atomic blasts are set off by the United States in different locations in the Marshall Islands. Russians explode first atomic bomb in 1949.	More uninformed medical research is done through the collection of blood and urine samples and medical exams. The United States expands bomb development due to fear of Russian bomb development.
1952	Russians explode their first hydrogen bombs. These bombs are believed to be stronger than any previous American bombs.	The United States ramps up the development of a hydrogen bomb in response to the Russians.
Early March 1954	The United States detonates a hydrogen bomb Bravo at Bikini. The bomb is the equivalent of 1,000 Hiroshima bombs. Radioactive "gritty white ash" falls on many places where Bikinians live.	After Bravo, U.S. scientists evacuate some islanders in Project 4.1 and resettle them on another contaminated island to further study radiation effects.
Late March 1954	Bikini, Rongerik, Utrik, Ujelang, and Likiep atolls are deemed off limits and people are moved again.	No forewarning was given to the Marshallese people of the impending blasts or subsequent blasts later that year.

(continued)

Table 3.1 (Continued)

1958	Bikinians sign an agreement with the United States turning over full-use rights of Bikini.	In return, Bikinians are given full-use rights to Jaluit Atoll and several other islands. They also would no longer be able to make claims against the United States concerning the Bikini Atoll.
1967–68	The United States considers returning Bikinians to their homeland because a study declared that well water at Bikini was now safe.	President Johnson promises that the Bikinians could now return to Bikini and build a "new and model community… with all possible dispatch."
1972	Three extended Bikinian families decide to move back to Bikini. Coconut trees had been planted on the island and some dwellings had been built.	The U.S. government withdraws its military personnel and equipment from Bikini and halts weekly air service to Kwajalein Atoll.
June 1975	U.S. Dept. of Interior, during regular radiation monitoring of Bikini, discovers that there are "higher levels of radioactivity than originally thought."	Another study by Brookhaven Laboratories contradicts the Interior Dept. and says that the readings are "probably not radiologically significant."
October 1975	Bikinians file lawsuit in U.S. federal court demanding a complete scientific study of Bikini and the northern Marshalls.	The United States agrees to conduct a more thorough study, but due to bureaucratic difficulties, it takes three years to begin.
1977	Further studies by independent scientists and the Department of Energy reveal high levels of radiation on Bikini.	The DOE tells residents of Bikini to eat only one coconut a day, and some food supplies begin to be flown in.

April 1978	Medical exams show the 139 residents of Bikini to be beyond U.S. permissible limits, and the Interior Dept. describes the radiation levels as "incredible." Plans are made, once again, to remove the population from Bikini. The lawsuit-forced scientific study begins, but only after evacuation has been completed.
1980s	Lawsuit is filed by Bikinians in 1981, but it is dismissed in 1987. Bikinians receive two trust funds for compensation because they had given up their lands for nuclear testing.
1994–2001	Bikinians file lawsuit against the United States for damages. Bikinians begin a tourism campaign to draw sport fishermen and scuba divers to visit the historic island.
2001	The Nuclear Claims Tribunal awards the Bikinians approximately $500 million. The tribunal cannot provide any of the award to the Bikinians because the tribunal was never given the money by the U.S. Congress.
2006–2009	Bikinians file lawsuits under the Fifth Amendment for the damages awarded in 2001. U.S. Court of Federal Claims grants motion by the United States to dismiss the suit in 2007. U.S. Court of Appeals affirms dismissal in 2009.
April 2010	Lawsuit is appealed to the U.S. Supreme Court in March: review is denied by the Court in July. The odyssey—and the work of the Bikinian people for just compensation under the Fifth Amendment—continues.
August 2010	UNESCO approves application for Bikini Atoll to become a World Heritage site. The World Heritage site will aim to, in Tomaki Juda's words, "remind all of us, around the world, of the need for global peace and the elimination of weapons of mass destruction."

NOTES

1. *The New York Times*, May 26, 1946.
2. J. Niedenthal, *For the Good*, 2.
3. D. Bradley, *No Place to Hide*, 50–51.
4. Ibid., 54–55.
5. Ibid., 93.
6. Ibid., 109.
7. Ibid., 108.
8. Ibid., 111.
9. Ibid., 88–89.
10. Ibid., 114.
11. J. Niedenthal, *For the Good*, 4.
12. E. Welsome, *Plutonium Files*, 194
13. Ibid., 194.
14. B. Johnston, *Half Lives*, 31.
15. Ibid., 31.
16. J. Niedenthal, *For the Good*, 4.
17. B. Johnston, *Half Lives*, 45–46.
18. J. Niedenthal, *For the Good*, 10.
19. Ibid., 12.
20. Ibid., 12.
21. Marshall Islands Consolidated Legislation, 2005.
22. Hon. Tomaki Juda's letter in N. Baker, *Bikini Atoll World Heritage*, 6–7. Nomination to include the Bikini Atoll on the World Heritage List was approved in 2010.

PART TWO

Using the Atomic Mindset: Native Americans, Guinea Pigs, and Uranium Cottage Industries

After a new technology is developed, it can lie fallow for a period of time, a time when the new invention is, for all intents and purposes, inert. Once it begins to be used, however, it not only begins to develop mindsets in most cases, but it also starts to affect human actions and, ultimately, the ethics of the new invention. That is, technologies, after being put into use, are no longer innocent. The use of atomic inventions is a prime example of this phenomenon. We have already seen how some uses of atomic inventions—luminescent paints, bombs, and uranium processing—began to develop a mindset of fascination and awe that was coupled with secrecy, deceit, and denial. In this second section, we will encounter atomic development as its various uses began to expand the mindset and take it into wider social, economic, health, and environmental venues.

Chapter 4 explores a distinctive aspect of the mindset: the different views that social groups can have with respect to the natural environment within and on the lands of the Navajo Indians of the American Southwest. As the title of the chapter suggests, there has been a clear tension between the dominant white and the Native Americans' values concerning the use of land. For the whites, the land of the Navajos is filled with a yellow rock—yellowcake—that holds the coveted uranium needed to create weapons of mass destruction and to power everything from submarines and spacecraft to electricity for our homes and workplaces. For the Navajo, as well as other native cultures, the yellow rock was a part of the land that should be left undisturbed and not used for human purposes. This tension between rocks and land, and the mining processes that do indeed disturb the land in ways injurious to both people and the land, tells the story of how mining, milling, transporting, and processing uranium has changed both the culture of the people who live on the land, and also the role that the government and atomic industries have played in keeping this tension at odds so that nuclear development would not be stalled.

The next two chapters are part of a piece investigating the myriad ways that humans have been treated as uninformed subjects of experimentation with radioactive materials. Chapter 5 begins with the decade-long legal and policymaking wrangling of government officials and scientists in the 1950s to enact a code of ethics for human subject rights, a code that, once approved, was deemed "top secret" due to its use of atomic terminology and, thus, was not used for several decades. The American Code, drawn almost verbatim from the Nuremberg Code developed after the Nazi Holocaust to insure the rights of human subjects on an international scale, was to lie dormant within the government top-secret archives and, in turn, allow uninformed experimentation to be conducted until the 1990s.

Chapter 6 continues the focus on *use* through a medley of three short vignettes. First, the story of the Oatmeal Boys takes place in the 1940s and 1950s at the Fernald School for Boys in Waltham, Massachusetts. For a decade, dozens of teenagers were fed oatmeal provided by Quaker Oats and administered through scientists from the Massachusetts Institute of Technology for the purpose of determining how well the nutrients "flowed through their bodies." The boys were never told of the nature of the oatmeal, nor the purposes of the experiments, until the 1990s. Second, the story of the Green Run centers on the mammoth atomic development facility in Hanford, Washington, in the late 1940s where plutonium was processed for use in the frenzied making of atomic bombs. The hurried process, due to fear that the Russians would develop bombs more quickly than the Americans, resulted in the release of enormous radioactive clouds that settled over thousands of acres of towns, pastures, and watersheds. Finally, the third story gleans from a plethora of events concerning the over-radiation of medical patients in cancer and other disease treatments. This vignette brings us right up to the present day and demonstrates how humans continue to be the subjects of questionable radioactive uses in a variety of contexts.

Chapter 7 concludes this second section on using the atom by telling the first part of a two-part story of how uranium milling became a cottage industry in the U.S. during the 1950s and 1960s. During this period, hundreds of private, independent contractors conducted top-secret uranium milling operations across the country. In these shops that were often located in abandoned manufacturing warehouses or other industrial sites, workers milled and drilled by hand uranium slugs that were then transported to the large uranium processing facilities that made up the backbone of atomic bomb development. This chapter homes in on one of these facilities that was, surprisingly, in an old warehouse residing in a residential area within the city of Oxford, Ohio, also the home of Miami University. The vignette draws upon archival documents such as top-secret contracts and instructional papers from that period. In addition, there are narratives from some of the men who worked in the milling shop that depict what it was like to work in this environment where they knew little of the nature of the work and the potential environmental and health effects associated with continuous exposure to radioactive materials and the disposal of the byproducts. The story sets the stage for what will follow in Part III when the existence of this facility was uncovered in the 1990s.

CHAPTER 4

Engagements with Rocks and Land: Uranium, Diné Culture, and the Yellow Monster

The Diné—the Navajo people of the American Southwest—have a deep connection with the land. The land is the medium through which they sustain their bodies, minds, and spirits. For the Diné, land, water, and living things are inseparable. In the words of one Diné woman, Rena Babbit Lane,

> my father had songs for the four sacred mountains, songs for performing the Beauty Way ceremony and water is one of the songs. When we are in harmony with the whole earth this is the Beauty way. But disrespecting the nature of life, mis-using water, fencing, capping off, bulldozing our water sources and giving our lands away causes dis-harmony with Nature and earth life. That is why the rain does not come anymore and why we have been suffering from a drought... That is why life is out of balance.[1]

Balance in the natural order of things, embedded in Diné and other indigenous cultures, serves as a central theme of origin stories. One prophetic Diné tale relates that when the Diné emerged into the fourth world—the present time—they were given a choice between two yellow powders: one a dust from the rocks and one the pollen of corn. The Diné chose the pollen, a choice that was blessed by the gods. The yellow dust of the rocks, however, came with a warning from the gods: If it is ever removed from the earth, then evil will come.[2]

For the non–Native atomic culture, the yellow rocks—yellowcake—contained uranium, and their desire was to remove it from the land as quickly and efficiently as possible. Thus, balance, for the non–Native atomic entrepreneurs, mining interests, and government personnel, wasn't found in the natural order but was located at the bottom of a bank ledger. Rocks and land: part and parcel of each other, but not so when perceived through the imperative of the atomic mindset.

It is no secret that the relationship between Native and non–Native cultures in the United States has been fraught with tensions, especially when it comes to disputes over lands and land treaties. Many treaties with Native Americans were never officially recognized by the U.S. government. When they were recognized, often they were just swept under the bureaucratic rug when it became convenient to do so. Prospecting for and mining uranium minerals on Native lands are classic examples of how the atomic mindset worked, abusing the land and its people to balance its ledger.

THE YELLOW ROCK AND NATIVE LANDS: A SYNOPTIC HISTORY

The problems associated with Native American lands and uranium exploration and mining actually began in the 1870s, long before the post–World War II rush. In 1871, for example, pitchblende was discovered in various areas of the United States from Connecticut to the Dakotas, Texas, and Colorado. Used in Europe for glazing and glass manufacture, this high-grade uranium ore was highly valued. The excavation of the ores in the late 1800s, however, paled in proportion to what would come in the twentieth century. In 1901, Madame Curie isolated the element radium found in the uranium ore, and the race was on to find more deposits. Radium became all the rage: First used in Europe as a curative for a variety of maladies, including cancer, it soon found its way into quackish cure-alls for everything from bladder and kidney problems, headaches, arthritis, and lung ailments to a host of sundry health problems.

This newfound miracle element, along with other metals such as vanadium, carnotite, and copper, pressured legislators to pass leasing rights laws to control Native American lands for mining. In 1902, Senate Bill 145 opened bidding on the Unitah Indian lands in Utah for rights to graze cattle and explore for uranium and other ores. The bidding for grazing and mining rights accelerated over the next two decades, as cattle ranching had grown many times over, and the growing interest in uranium—thanks to the discovery of radium as a

potential cure-all and base for the production of paints like Undark—
sparked an entrepreneurial fever. Further, vanadium was found to
increase the tensile strength of steel, so plenty of eastern U.S. money
was available to aid in the exploration of this and other precious industrial metals. As Kevin Phillips notes in *Democracy and Wealth: A Political History of the American Rich*, of the 22 millionaires residing in the
U.S. Senate in 1902 to 1903, seven had made their wealth through
mining. The connection between mining money and political power
was well embedded in the American pursuit of rocks.[3]

At work here, too, was the all-powerful concept of Manifest
Destiny—that the white race was the true owner of anything that
slaked the unquenchable thirst for progress. As William Jones, the
non–Native American commissioner of the Bureau of Indian Affairs,
said most pointedly in arguing in front of the 1902 Senate panel,

> I do not believe in reserving large tracts of land for the exclusive
> use of Indians. I believe it ought to be thrown open as rapidly as
> possible. I understand, though, that there is a treaty arrangement
> with those Indians that makes it necessary for you to treat with
> them before it can be thrown open to settlement . . . and I think
> to get them to agree to open the reservation you have got to use
> some means to open the land.[4]

And, put in even more raw, biased terms, Senator Joseph Rawlins of
Utah argued that, "If any money is given to them they gamble it away
and dispense with it. It is not in the best interest of the Indians to pay
them money . . . These are Indians who can not intelligently deal with
this subject independently."[5]

Despite the push in Congress to gain more government control
over the lands during the next 17 years, no bills were passed to make
it so. Politicians were distracted by international strife and war; and
on the domestic front, corporations and ranchers, believing that
government regulations would interfere with their unfettered access
to profit and land, kept still. Thus, while the political gaze turned elsewhere, the Native lands issue lay fallow until 1919. Then, during
annual appropriations discussions by Congress for the Bureau of
Indian Affairs, an amendment opened up virtually all western lands
to prospecting and mining—lands, according to the amendment, that
had been "heretofore withdrawn from entry under the mining laws
for the purpose of mining for deposits of gold, silver, copper and other
valuable metalliferous minerals."[6] Now anyone could enter Indian

lands without Interior Department permission and begin poking
around for evidence of metalliferous minerals. Under this amendment, should a prospector locate a potential lode, he would pay
$1 per year to the federal government for up to 40 acres and, as ore
was removed from the land, he would pay a royalty of 5 percent of
the net value of the mineral (the amount left after all mining operation
expenses had been subtracted).[7]

Rocks were again in high demand, but the market would soon
wither for a variety of reasons. Chief among them was the discovery
in 1913, six years prior to the 1919 legislation, of the world's richest
uranium reserves in the Belgian Congo (now Democratic Republic
of the Congo). These pitchblende ores contained about 70 percent
uranium, roughly 35 times the amount found in the rocks of the
southwestern United States. The discovery of Belgian Congo ore
was held in secret, like so many things atomic, but this time by the
Belgian government. Belgium wanted to dominate this burgeoning
market while the demand for radium by makers of luminous paints
and medical treatments exploded. However, this secret was uncloaked
in the 1920s, and the eyes of uranium markets quickly shifted to Afri-
ca's high-grade ore and away from the yellowcake of the Native
American lands. For another 20 years, until Einstein's letter and the
idea of an atomic bomb were unleashed, the southwestern desert was
more or less left alone when it came to radioactive hunting. All of this
would soon change.

TRINITY AND LEETSO: THE PROPHECY OF THE YELLOW ROCKS COMES TRUE

Leetso, the monster that the Diné believed resided in the yellowcake dust, was born at 5:30 a.m. on July 16, 1945, at Alamogordo,
New Mexico. The Diné, however, were unaware that the monster of
the yellow rock had arrived. The pregnancy had been long—more
than three years—and the labor had been intense for the Hanford,
Los Alamos, Oak Ridge, Chicago, and other Manhattan Project mid-
wives. The yellow rock of Mother Earth had grown within the great
industrial complex's womb into weapons-grade plutonium and was
given birth on that mid-July morning in a ball of fire that turned the
desert into glass. This was the first full-scale atomic blast ever
attempted and, in a fitting irony, was dubbed Trinity, thus invoking
another culture's origin story.[8]

The first atomic bomb blast 1/40th second after explosion 5:50 a.m., July 16, 1945, Trinity Site, Alamogordo Air Base, New Mexico. (Photograph by Berlyn Brixner. Photo provided by the Defense Threat Reduction Agency (DTRA).)

Berlyn Brixner, the official photographer of the birthing, described the moment as a "tremendously brilliant light that comes out of the initial part of the explosion, then the formation of a ball of fire, which becomes an immense ball of white-hot material. That's the image that persists in my mind."[9]

The moment inspired the Manhattan Project physicist and Trinity explosion observer Robert Oppenheimer to wax philosophical:

> We knew the world would not be the same...Most people were silent. I remembered the line from the Hindu scripture, the Bhagavad-Gita: Vishnu is trying to persuade the prince that he should do his duty and to impress him he takes on his multi-armed form and says, "Now I am become Death, the destroyer of worlds." I suppose we all thought that, one way or another.[10]

For another observer, the scientist Isidor Rabi, the Trinity moment was truly a birth ripe with portent: "A new thing had just been born; a new control; a new understanding of man, which man had acquired over nature."[11]

Leetso would now travel halfway around the globe to visit Hiroshima and Nagasaki, unleashing the monster's fury and becoming Death, the destroyer of worlds. But most of the Diné tribal members did not know that their yellow rock monster was the source of the world-changing explosions in Japan. Even though the Diné had answered the call of duty and generously enlisted in the war effort—at higher percentages than the general American population—they were unaware of the role that their lands had played in the bomb's development. "Traditional Navajos would have been horrified had they known what others would do with their yellow dirt," note Navajo writers Esther Yazzie-Lewis and Jim Zion in "*Leetso*, The Powerful Yellow Monster."[12] Had they known, they likely would have balked at entering the mines. As it was, the atomic mindset was just too powerful and its secrecy too pervasive. So enter Leetso's lair they did.

PROSPECTING, MINING, AND BEING AN AMERICAN CITIZEN IN THE DINÉ WORLD

Prospecting is just the tip of the uranium and atomic development industry iceberg, merely initiating the atomic development cycle. It is literally a scratching of the surface in that seeking ore most often imposes a relatively minor disturbance to the land (although it can entail excavation). Once ore is found, it needs to be mined using extraction procedures that involve digging, tunneling, scraping, and exploding the land with dynamite and other explosives. Sometimes extraction means forcing caustic chemicals into the land and drawing the ore out into processing units that "wash" the ore from the chemical soup. Following the mining, the ore must be further processed, refined, hauled, treated again, and then made into the various radioactive elements and products used by the atomic and nuclear industries. Finally, the remnants of the mining, the trucking, the processing, and the waste materials must be disposed of.

For the most part, Diné miners worked in the early stages of prospecting and mining. As far as prospecting was concerned, the Diné certainly had a few notables, but only a few. One, Paddy Martinez, is a prime example of the Native American prospector or at least prospector's aide. Martinez, a sheepherder by profession, was something of an accidental prospector. He didn't have sophisticated equipment, but, having wandered the land around Grants, New Mexico, for many years, he knew it intimately. One day in 1951, while riding on horseback, Martinez spied what he said was a "yellow rock" exposed under some large boulders. He knew some white prospectors were searching for such rocks, so he dutifully reported to them his find. The land, as it turned out, was owned by the Santa Fe Pacific Railway. The find was taken over by private mining interests, resulting in at least five major mines over the ensuing years. Martinez received a $250 finder's fee.

Martinez is but one instance of an irony that pervaded the Diné culture's role in atomic development: He was a Diné and, by definition, was a man who loved the land and the intimate connection the Diné had to it. But he also was an American patriot, as were many Diné, and was inclined to answer the call to duty, even if that call was duty to the U.S. government or business interests. A prime example of this patriotic spirit is the Navajo "code talkers" of World War II. The American intelligence services had been perplexed in creating a secret code that could not be broken by Japanese and German intelligence. The answer to the American code writers' dilemma was discovered in the Diné language. "Navajo answered the military requirement for an undecipherable code because Navajo is an unwritten language of extreme complexity. Its syntax and tonal qualities, not to mention dialects, make it unintelligible to anyone without extensive exposure and training. It has no alphabet or symbols and is spoken only on the Navajo lands of the American Southwest. One estimate indicates that fewer than 30 non–Navajos, none of them Japanese, could understand the language at the outbreak of World War II."[13] The code, written and translated by the Diné for the American military, proved to be unbreakable and is believed to have been a significant influence in the ending of the war. Praise for their skill, speed, and accuracy accrued throughout the war. For example, "Major Howard Connor, 5th Marine Division signal officer, declared, 'Were it not for the Navajos, the Marines would never have taken Iwo Jima.' Connor had six Navajo code talkers working around the clock during the first two days of the battle. Those six sent and received over 800 messages, all without error."[14]

The code talkers were willing to give two of their most precious possessions—their language and bodies—to create these unbreakable secret codes for use by the American military. Paddy Martinez and Diné miners gave their bodies and land. Diné walked a fine tightrope through the atomic mindset. On one side of the line was their loyalty to their American compatriots. On the other was a mindset rooted in their culture and land—land that eventually would be taken right out from under their noses.

RADIOACTIVITY, HEALTH, AND "DOG HOLES"

Working in the uranium mines on Diné land was arduous, dangerous, and unhealthy for the Diné miners. "Inside the interior of the mine
was a nasty place, smoky, especially after the dynamite explodes," testified Diné miner George Kelly to a U.S. Senate committee in 1979.

We run out of the mine and spend five minutes here and there
and were chased back in to remove the dirt by hand in little train
carts. . . . The water inside the mine was used as drinking water,
but no air ventilation, however. The air ventilators were used
only when the mine inspectors came and after the mine inspectors leave the air ventilators were shut off.[15]

Another Diné miner at the same Senate hearing said that,

[W]hen I went to work [in 1969], I was never told anything inside
the mine would be hazardous to my health later. It really surprised us to find out after so many years that it would turn out like
this, that it would kill a lot of people. They said nothing about
radiation or safety, things like that. We had no idea at all.[16]

Uranium "dog hole" mine with wood supports. (Photograph by Murray
Haynes. From *If You Poison Us*, (c) 1994 Peter H. Eichstaedt, used by permission of Museum of New Mexico Press.)

Uranium "dog hole" mine on Navajo land. (Photograph by Murray Haynes. From *If You Poison Us*, (c) 1994 Peter H. Eichstaedt, used by permission of Museum of New Mexico Press.)

Mining is always dangerous, and there have been significant strides made in contemporary times to protect miners. For instance, in 2008, U.S. mining of all types had become statistically one of the safest of high-risk jobs.[17] But such statistics elide the issues of long-term health effects. In West Virginia and Pennsylvania coal mines and on the Gulf Coast oil rigs, safety and long-term health and environmental effects are still a huge problem today. In these recent examples, once again, corporate oversight and government regulations go by the wayside when it comes to turning higher profits. Nevertheless, mining and other forms of natural resource extraction have become more attuned to the issues of health and safety, advancing beyond canaries.

Such was not the case in the Diné mines of the 1950s through the 1970s, where the mining conditions were crude and dangerous. Some of these uranium mines were sometimes no more than "dog holes"[18] consisting of small openings with absolutely no ventilation for dust, smoke, and radon gas. Some of the mines were crude dugouts under precarious rock formations, supported by timbers.[19]

Yet from these crude mines—and there were hundreds of them across the Navajo lands as well as in Canada and Alaska—came tremendous amounts of rock containing Leetso. The rocks so loved by the corporations and governmental agencies intent on developing the atomic mindset were being scavenged by the millions of tons, all to produce a few tons of plutonium over a 20-year period of time.

The Atomic Energy Commission's guaranteed prices for uranium and the bonuses for new deposits would continue through the 1950s, but by the late 1960s, the boom was more or less over. Uranium stockpiles had grown to the point that uranium ore was hardly worth the cost of extraction, and the AEC estimated that we had enough raw ore to last well into the twenty-first century. The incentive for prospecting was gone, and so were most of the prospectors who had packed their bags and Geiger counters and driven their Jeeps back to wherever they had come from. However, some of the participants in this get-rich-quick scheme (and purported patriotic duty) of finding uranium stayed put. The Native Americans of the Colorado Plateau, the Alaskan ranges, and a few other spots stayed at home on their land. But on this homeland remained the residue from the monumental extraction enterprise, and it was piling up, literally, at Church Rock, New Mexico. The radioactive residue and its shocking flood on July 16, 1979, would eventually find its way to Washington, DC, and the chambers of the U.S. Congress.

FROM CHURCH ROCK TO CONGRESS: LEETSO GOES TO WASHINGTON[20]

During the night of July 16, 1979—exactly on the 34th anniversary of Trinity—nearly 100 million gallons of water carrying 1,100 tons of radioactive material coursed silently through the desert of Church Rock, New Mexico, toward the Rio Puerco, a modest desert river that provides water for the Navajo people, their livestock, and crops who live along it. Instead of its customary trickle of untainted water, the Rio Puerco now held a torrent of radioactive water—with a pH of 1, about the same as battery acid[21]—that carried thorium, radium, lead, polonium, and sundry other radioactive elements and heavy metals. The source was a holding pond at a uranium mining and milling facility owned by United Nuclear Corporation (UNC). This mine was the largest operating underground uranium mine in the United States. It employed about 800 people and the adjoining mill another 150. On that night, this large facility became, apart from nuclear arms testing, the site of the "largest release of radioactive materials in the continental United States."[22]

The flood came in the darkness unseen, its origins not immediately known to the Navajo who lived on this land. The 350 Navajo families that lived along the river were surprised when all of this water suddenly appeared in the early morning hours. "There were no clouds, but all of a sudden the water came," remembered Rio Puerco resident Herbert Morgan. "I was wondering where it came from. Not for a few days were we told."[23]

Eventually, the story of Church Rock would be told to a national audience, but there was a three-month gap before this would happen. For some reason, this gargantuan spill of radioactive materials in the desert Southwest did not command the immediate attention that another, lesser nuclear incident drew earlier that year at Three Mile Island, Pennsylvania, on March 28. The partial meltdown of one reactor core at the Three Mile Island nuclear power plant drew much immediate media coverage, and a presidential commission was formed to study the accident within two weeks of that event.

Nevertheless, the Church Rock spill did receive a Congressional hearing. On October 22, 1979, a four-hour hearing was held that included testimony by several Diné tribal officials, as well as representatives of United Nuclear Corporation. The hearings began with Rep. Morris K. Udall, the chair of the Committee on Interior and Insular Affairs and Subcommittee on Energy and the Environment, providing a synopsis of what had happened on that July morning:

> On July 16 of this year a uranium mill tailings impoundment failed, releasing 93 million gallons of contaminated liquid and 1,100 tons of radioactive waste into an arroyo near Church Rock, NM. The radioactive and chemically dangerous materials were carried to the Rio Puerco, through Navajo grazing lands ... leaving contaminated residue over a distance of close to 100 miles. This morning we will hear of the social disruption and hardship this accident has caused the Navajo people and of the economic difficulties it has created for both the United Nuclear Corp. and for its employees.[24]

Rep. Udall continued by laying out the committee's scope of jurisdiction and, in a most startling comment, suggested that the long-term radioactive contamination brought by the flood was to remain in the land forever.

> Our concern in these matters derives from several areas of the Interior Committee's jurisdiction: dam safety, Indian affairs, and finally regulation of the nuclear industry. It is likely the levels of contamination will not result in the loss of human life if the spill is properly cleaned up. It is important to note, however, that all of the contaminated material will never be completely removed from the environment.[25]

He then drew the introductory remarks to a close with yet another telling characteristic of the atomic mindset: the ambivalence with which government and industry contractors and regulatory agencies sometimes do their job.

> In the case of the Church Rock tailings site, at least three and possibly more Federal and State regulatory agencies had ample opportunity to conclude such an accident was likely to occur. Yet there is an indication that none of the regulatory authorities required detailed independent assessments of the company's construction practices. The cracks, which eventually led to the failure, began to appear in December of 1977. There is some question whether the company was monitored to assure proper mitigating measures were being undertaken.[26]

It was revealed later in the hearings that the dam had been improperly constructed atop "areas of shallow and deep bedrock. The unusual configuration of the bedrock acted like a fulcrum and caused the cracking."[27] Thus, the faulty dam construction led to the eventual breach that resulted in the torrent of radioactive water.

David Hann, the chief operating officer of UNC, denied the claims by the Army Corps of Engineers and the New Mexico State Engineer's Office that the breach was a result of poor engineering and construction. Instead, he made the argument that UNC had done everything in its power to make sure the tailings dam was safe. Besides, he claimed, "It is the opinion of our professional staff that there was no substantial radiological danger created by the mill."[28] In his view, the radioactive effluent was essentially benign. "The tailings liquid in a facility such as Church Rock is a dilute acid. If a person fell into our tailings pond, the water would not taste good and his eyes would smart, but his health would not be endangered . . . the tailings did not present a health hazard."[29] He added that building the dam was just a price of doing business: "The breach was like many things you undertake, they have a risk and we undertook this."[30]

The Diné members at the hearing saw things in quite a different light, especially when it came to potential health hazards. Frank Paul, the vice chairman of the Navajo Tribal Council, explained in no uncertain terms that the Diné wanted to know more about the threat that the spill presented, and they wanted to be involved in the decisions concerning future mining and milling.

> As a result of lack of involvement in mining and milling operations by the Navajo Nation, we have not [been] advised as to the potential dangers of radiation on Navajo workers and their families. Hundreds, if not thousands of Navajo uranium miners are contaminated from the dust and air in the mines commonly called dog holes. Navajo families used scrap rock that was available in the uranium mines to build homes. Further, when mines are abandoned, they are simply left as they had been on the last day of mining. In general, the uranium mining industry within the Navajo Nation proceeded without any regard for the health and safety of the Navajo people. The Navajo Nation was taken for granted as some kind of proving ground or national sacrifice area.[31]

Helen George, a member of the Navajo Community Action Committee, added to the litany of everyday problems the spill had perpetrated on the local community. "The tailings spill of July 1979 has impacted the community so that all affected communities have been sent reeling with uncertainty, confusion, and bitterness," she said.[32]

> This incident has only led to distrust of uranium development activities in people's backyards, as is happening all over Navajo occupied and owned land. In the people's efforts to deal with the tailing spill, we have only been told that this incident does not warrant an emergency, and people have tried to get assistance, to determine the type and magnitude of contamination, but we have not been told how long we cannot wash and whether the livestock are contaminated.[33]

She also indicated that the contamination had been so widespread that all of their food sources had been affected. "We have no alternative source of food, so we have requested food stamp assistance, but this has been denied."[34] She ended her brief comments by asking a most common question by people who are confronted with the after effects of Leetso. "What are the Navajo people to do?"[35] she queried. And then, echoing the long-held and deep belief of their connection to the earth, she reminded the congressional members of that intimate connection.

> We are very much concerned about what has happened, because for centuries, ever since people emerged out of the Earth, the Earth has nourished us and given us life. Jobs programs come and go, food stamp programs come and go, but the earth remains, and this is where people get their life from.[36]

The hearing ended with a return to the sense of Udall's opening remarks about the permanent nature of the radioactive debris, but this time the comments came from Paul Robinson, an environmental analyst for the Southwest Research and Information Center. He stated that,

> As a result of the first month of contamination, I believe the spill is uncleanable. There has been very fast ground water movement as a result of the spill and this has a great concern for those following the spill in detail. About 3,100 tons have been cleaned to date, or less than one percent of the total spill. The cleanup of less than one percent means that the radioactive materials are moving through the Rio Puerco system into Arizona, heading toward Lake Mead and the Little Colorado. The tailings material has a pH of 1, very highly acidic. It is not safe material.[37]

The hearing concluded abruptly after Robinson's comments on the widespread effects of the Church Rock spill. Following the hearing, there were more studies done by state and federal agencies concerning the levels of radioactive contamination, but the government officials allowed UNC to reopen the mine and mill in less than five months. The same tailings pond was used. In 1981, Paul Robinson told a reporter in an interview that some changes had been made to the dam but that "constant seepage up to 80,000 gallons of contaminated liquid per day became a mainstay."[38]

The fact that radioactive liquid continued to seep into the Rio Puerco and other streams, not to mention the ground water, prompted the federal government to have UNC post warning signs saying, "Contaminated Wash. Keep Out!" Ironically for the Diné—who had given the American government the Diné language for the code talkers of World War II—the signs were posted in English. As Tom Charlie, a Navajo sheepherder explained,

> Most of us can't read, write or speak English. The signs do no good. If neighbors know we are from the Rio Puerco wash, they won't shake our hands. They ran from me. . . . It is wet now, but on days when it dries up, the wind will come along. The dust settles on the grass. The sheep eat it. We eat the sheep. We wonder what that does to our lives.[39]

Thus, the cycle of rocks and land remains unbroken, except for the Diné of the Puerco River, where the cycle is ruled by Leetso.

NOTES

1. (lady n.d.).
2. P. Eichstaedt, *If You Poison Us*, 47.
3. K. Phillips, *Democracy and Wealth*, 240.
4. P. Eichstaedt, *If You Poison Us*, 15.
5. Ibid., 16.
6. Ibid., 19.
7. Ibid.
8. Brixner Photo #99. Trinity, the first atomic bomb blast.
9. R. Del Tredici, *At Work*, 186.
10. R. Rhodes, *Making*, 676.
11. Ibid., 672.
12. E. Yazzie-Lewis, *The Navajo People*, 3.
13. Naval History & Heritage Command: Navajo Code Talkers: World War II Fact Sheet, http://www.history.navy.mil/faqs/faq61-2.htm.
14. Ibid.
15. B. R. Johnston et al., *Half Lives*, 102.
16. Ibid., 103.
17. G. Robinson, 2010, Michigan Technological University's Seaman Mineral Museum, personal conversation.
18. P. Eichstaedt, *If You Poison Us*, 81.
19. Ibid., 82.
20. Congressional Subcommittee on Energy and the Environment, 1979.
21. Robinson in Bryant and Mohai, *Race and the Incidence of Environmental Hazards*, 158.

22. H. A. Saleem, in B. R. Johnston et al., *Half Lives*, 104.
23. H. Wasserman and N. Solomon, *Killing Our Own*, 177.
24. Congressional Subcommittee, 1.
25. Ibid., 2.
26. Ibid., 3.
27. Ibid., 22.
28. Ibid., 24.
29. Ibid., 30.
30. Ibid., 29.
31. Ibid., 7.
32. Ibid., 14.
33. Ibid., 14.
34. Ibid., 14.
35. Ibid., 14.
36. Ibid., 14.
37. Ibid., 47.
38. Wasserman, *Killing Our Own*, 182.
39. Ibid., 182.

CHAPTER 5

Dangerous Familiars Part 1: Nuclear Science and Its Human Subjects

There is no safe level of radiation exposure. So, the question is not: What is a safe level? The question is: How great is the risk?
—Dr. Karl Z. Morgan, Director of Health Physics, Oak Ridge National Laboratories, 1943–1972[1]

In mid-April 1946, a U.S. military plane crossing high above the South Pacific transported soldiers home from the war with tender cargo. The soldiers had been joined by a young Australian mother, Freda Shaw, and her four-year-old son, Simeon. Simeon, called Simmy by his family, and Freda had boarded the plane in Sydney for the long flight to San Francisco. The Red Cross had flagged this a mercy flight because Simeon was sick. He was suffering from a large and painful tumor in his leg caused by a virulent form of cancer—osteogenic sarcoma.

An Australian doctor, whose identity is still unknown because the name was struck from the records by the Department of Energy years later, had learned that doctors at the University of California Hospital in San Francisco (UCHSF) might be of assistance in curing Simmy's apparently incurable condition. Curiously, it appears that Simmy's Australian doctor or some other medical official in Australia had studied the work of Dr. Joseph Hamilton. Hamilton, a medical researcher at UCHSF who was simultaneously employed by the Manhattan Project, had been conducting secret radiation experiments for some time, and now that the war had ended with the dropping of the atom bombs, he had found another venue for experiments on humans:

UCHSF. In a memo dated just seven months prior to Simmy's visit, Hamilton described a project he hoped to pursue.

> The next human subject that is available is to be given, along with plutonium 238, small quantities of radio-yttrium, radioactive strontium and radio-cerium. The procedure has in mind two purposes. First, the opportunity will be presented to compare in man the behavior of these three representative long-lived fission products with their metabolic properties in the rat, and second, a comparison can be made of the differences in their behavior from that of plutonium.[2]

To the eyes and hearts of Simmy's family, the flight to San Francisco was a last-ditch effort to help the boy. Freda had faith in the mysterious medical procedures and hoped-for cure awaiting her son, but they were unknowns. Upon arrival in California, Freda said she was hopeful "because I have to be."[3] Indeed, hope for a cure was literally all she had at this point. Hamilton had hopes, too. His hopes were for information. Simmy was destined to be a human guinea pig, or maybe more accurately, a "rat."

Freda and Simmy were met in San Francisco by a media blitz celebrating the mercy flight's role in helping to manifest a cancer treatment miracle. Newspapers around the country carried headlines that used words like *hope* and *mercy*. Reporters with cameras and notebooks closing around Simmy at the air base captured for public consumption the frail and feverish little boy who was now in considerable pain.

Simmy was taken immediately to the university hospital, where he was thoroughly examined by doctors; and then he was separated from Freda, who was allowed to see him only three times a week. Over that next week, Simmy was treated for ear infections, provided with antibiotics and painkillers, and made ready for the inevitable injections that might heal his withering body. On April 26, he was injected with a radioactive cocktail consisting of three isotopes: cerium, plutonium-239, and yttrium (although some evidence purports that it was actually rubidium). The dose was about 50 times the radiation an average person might receive over 50 years. Simmy was also given a new name by the doctors—CAL-2—a name that would be his anonymous identifier concealed in top-secret files for the next 50 years. He received a second injection a few days later, but the contents of that injection are not documented.[4]

Simmy remained in the hospital for about a month so doctors could examine his x-rays and monitor changes to the tumor in his leg. Codeine and aspirin failed to stabilize his temperature fluctuations or diminish his pain; his temperature spiked as high as 104 degrees, and he continued to suffer considerable pain. At one point, the doctors decided to conduct a biopsy of the tumor. Upon receiving written consent from Freda to put Simmy under anesthesia—the only medical consent she was asked to sign during her time in San Francisco—surgeons removed portions of the tumor along with a section of bone and some muscle tissue. Simmy returned to his bed in the hospital ward, where he remained until released on May 25.

Sailing from San Francisco on June 14, Simmy and Freda took an entire month to return to Sydney. Simmy remained in a Sydney apartment with his family until January, never returning to his hometown of Dubbo. On January 6, 1947, Simmy succumbed to his excruciating disease. His brother Joshua remarked that when he last visited his brother's room, "Simmy was screaming with pain. I couldn't stand it."[5] Simmy, or more accurately CAL-2, would remain an anonymous human subject of radiation experimentation well into the 1980s. In 1997, his survivors were given $262,500 in an out-of-court settlement with the hospital—one of the smallest awards given, and given without the hospital admitting any official blame. Joshua angrily remarked at the time, "Morality has to be accounted for. The Nazis tattooed numbers on their victims' arms. The American government gave Simeon Shaw the number 'CAL-2.' "[6]

HUMAN SUBJECT TESTING
AND INFORMED CONSENT

Scientific human testing can be widely categorized into informed and uninformed. The advent of informed testing traces back to a famous early twentieth-century legal challenge. *Schloendorff vs. The Society of the New York Hospital* brought to light a case of surgery performed on an uninformed patient. Mary Schloendorff had entered the New York Hospital in January 1908 "suffering from some disorder of the stomach."[7] Mary agreed to be examined while under anesthesia. A surgeon detected a fibroid tumor while she was sedated and, without Mary's knowledge, removed the tumor. Following the procedure, she developed an infection and gangrene, resulting in the amputation of several fingers. She sued the hospital, and after two appeals, the court ruled in favor of Schloendorff. In final comments, Justice Cardozo of the New York Court of Appeals stated, "In the case at hand, the wrong complained of is not merely negligence. It is trespass. Every human being of adult years and sound mind has the right to determine what shall be done with his body."[8] Over the next decades, the precedent set by this ruling would enforce informed consent in the medical and scientific research communities—that is, until informed consent met the atomic mindset.

Uninformed testing, as might be suspected, includes studies in which the human subject knows little or nothing about the actual nature of the testing being conducted. This, arguably, is potentially the most unethical of human testing types, but it nevertheless has been done in all kinds of ways over the centuries. In the twentieth century, uninformed testing gained notoriety during the horrors of the Nazi death camps and on American soil through 40 years of Tuskegee experiments to observe the effects of syphilis as it progressed through the bodies of unsuspecting African-American sharecroppers.

Adopted after the post–World War II war crimes trials, the Nuremberg Code was intended to comprehensively regulate how human subject research should be conducted throughout the course of any project. The 10 elements of the code required that any experiments on humans, be they physical or psychological in nature, must include information. Human subjects must understand the purposes, methods, and potential risks involved. Furthermore, the subject has the right to end the experiment if he or she so wishes. The code makes it explicit that such experiments "should be such as to yield fruitful results for the good of society, unprocurable by other methods or means of study, and not random and unnecessary in nature."[9] The code may appear to be clear and straightforward; one would expect that such open and reasonable guidelines would have been equally expedient to implement by scientists, doctors, and other specialists conducting radiation studies on human subjects. In particular, one would imagine that the immediate link between the code and the Holocaust would be a powerful deterrent. Who would want their study to be associated with the most publicized and horrendous crimes known to humankind? Would the code not, at the very least, amplify the wish to *not* do unto others that which we would *not* want done unto ourselves?

Nuremburg Code infractions—as well as infractions to the Golden Rule—became commonplace and entirely acceptable to the atomic mindset in the years and decades following the Nuremberg trials.

NUREMBERG CODE

1. The voluntary consent of the human subject is absolutely essential. This means that the person involved should have legal capacity to give consent; should be so situated as to be able to exercise free power of choice, without the intervention of any element of force, fraud, deceit, duress, over-reaching, or other ulterior form of constraint or coercion; and should have sufficient knowledge and comprehension of the elements of the subject matter involved as to enable him to make an understanding and enlightened decision. This latter element requires that before the acceptance of an affirmative decision by the experimental subject there should be made known to him the nature, duration, and purpose of the experiment; the method and means by which it is to be conducted; all inconveniences and hazards reasonable to be expected; and the effects upon his health or person which may possibly come from his participation in the experiment. The duty and responsibility for ascertaining the quality of the consent rests upon each individual who initiates, directs or engages in the experiment. It is a personal duty and responsibility which may not be delegated to another with impunity.

2. The experiment should be such as to yield fruitful results for the good of society, unprocurable by other methods or means of study, and not random and unnecessary in nature.

3. The experiment should be so designed and based on the results of animal experimentation and a knowledge of the natural history of the disease or other problem under study that the anticipated results will justify the performance of the experiment.

4. The experiment should be so conducted as to avoid all unnecessary physical and mental suffering and injury.

5. No experiment should be conducted where there is an a priori reason to believe that death or disabling injury will occur; except, perhaps, in those experiments where the experimental physicians also serve as subjects.

6. The degree of risk to be taken should never exceed that determined by the humanitarian importance of the problem to be solved by the experiment.

7. Proper preparations should be made and adequate facilities provided to protect the experimental subject against even remote possibilities of injury, disability, or death.

8. The experiment should be conducted only by scientifically qualified persons. The highest degree of skill and care should be required through all stages of the experiment of those who conduct or engage in the experiment.

9. During the course of the experiment the human subject should be at liberty to bring the experiment to an end if he has reached the physical or mental state where continuation of the experiment seems to him to be impossible.

10. During the course of the experiment the scientist in charge must be prepared to terminate the experiment at any stage, if he has probable cause to believe, in the exercise of the good faith, superior skill and careful judgment required of him that a continuation of the experiment is likely to result in injury, disability, or death to the experimental subject.

Source: Reprinted from *Trials of War Criminals before the Nuremberg Military Tribunals under Control Council Law* 2(10): 181–182. Washington, DC: U.S. Government Printing Office, 1949.

Ironically, the U.S. government and military adopted the code but kept it secret, thus rendering it impotent. Secrecy kept the code out of sight, but not out of the atomic mindset.

TOP SECRET: THE HUMAN SUBJECTS CODE OF ETHICS

Many scientists, military personnel, and politicians involved with atomic development were clearly concerned after World War II with how humans should be treated in any experiments regarding radiation. They were caught in a double bind. On the one hand, they did not wish to cause harm to humans, and on the other hand, they knew some harm was inevitable. However, the drive to find out more about how radiation affected humans was paramount. And guilt among some atomic scientists drove them to find reconciliation in life-affirming applications for their knowledge.[10] Many researchers, however, were willing to conduct their research knowing that test subjects would, indeed, be harmed.

Dr. Joseph Hamilton, the same doctor so influential in the design of treatments such as Simmy Shaw's, wrote a memo in November 1950 addressed to Dr. Shields Warren, head of the Atomic Energy Commission's Division of Biology and Medicine. In the memo, Hamilton calls for studies in radiation effects. Hamilton knew such research was essential for developing the means to protect troops and civilians from atomic attacks on the United States or its allies, as well as from friendly radioactivity released during bomb testing. Hamilton led the march toward more radiation testing and would continue to be an advocate for the next decade —until he succumbed to cancer resulting from radiation exposure.

Despite his zealous pursuit of atomic knowledge, Hamilton also identified a fly in the ointment. "For both politic and scientific reasons," he wrote, "I think it would be advantageous to secure what data can be obtained by testing large monkeys or chimpanzees. If this is to be done in humans, I feel that those concerned in the Atomic Energy Commission would be subject to considerable criticism as admittedly this would have a little of the Buchenwald touch."[11] Alluding to "the Buchenwald touch," of course, indicates Hamilton's desire to not be associated with a scientific project that sent thousands of testing subjects to their deaths after having been injected with various biological agents, such as typhus bacteria. This dilemma of acquiring scientific data and at the same time protecting human subjects was rapidly becoming an issue of great concern for the military, scientific, and medical communities. Ironically, a resolution to this dilemma came from neither of those communities. Instead, the solution came through the work of a civilian Pentagon employee and the ex-head of the world's largest automobile manufacturer.

In October 1952, Dr. Melvin Casberg of the Armed Forces Medical Policy Council decreed in a classified memo, "It was resolved that the ten rules promulgated at the Nuremberg Trials be adopted as the guiding principles to be followed" when experimenting on humans.[12] This decree would draw the code out of the shadows of secrecy and begin the process that would eventually lead to the adoption of the Nuremberg guidelines by the armed services and the U.S. government. However, en route to the final approval, two things happened, causing controversy and, ironically, a strengthening of the code. The first event involved Anna Rosenberg, a strong-willed and influential civilian Pentagon official. Rosenberg had a well-earned reputation for being a tough bargainer. She also understood labor law. Rosenberg, formerly employed in New York as a labor policy developer, knew what was important in constructing documents that affected both management and workers. She understood the legal significance and consequences of attaining a personal signature on a binding document.

Just days after Casberg's memo calling for the code, Anna Rosenberg advised the council's lawyer, Stephen Jackson, to amend the code by adding what she knew would be essential wording. Her recommendation for amending Rule #1 specified acquiring human subjects' signatures. Going further, she emphasized that this signature be witnessed by at least one other person. This recommendation was eventually adopted, but not until Pentagon officials and medical advisors raised a host of objections. Most pointedly, these military and medical research personnel did not want any code to interfere with what they believed was the integrity of scientific research. How, they wondered, could scientific objectivity be upheld when subjects knew the reasons for the studies? One might wonder about the likelihood of recruiting volunteers to willingly become guinea pigs. And—most critical to the scientific and military atomic mindset—how would secrecy about atomic development be assured? The cat would be out of the bag.

Despite the questions and strong objections made by medical and military officials, the revised code was sent up the ladder for final approval. By the time this occurred in January 1953, newly elected President Dwight Eisenhower had nominated Charles E. Wilson for Secretary of Defense. Wilson came to this post after serving as chairman of General Motors, where he had coined the famous phrase, "What is good for General Motors is good for America!" Although Washington insiders didn't think Wilson knew much about national political intricacies, Wilson was indeed quite knowledgeable about labor management issues and thus, like Anna Rosenberg, he understood the need for care in developing policies that affected the rights of workers. As Jonathan D. Moreno suggests in his book *Undue Risk: Secret State Experiments on Humans*, the requirement that a subject's signature be included in the new code "would have been consistent with the attitude of a former CEO of an automobile manufacturing giant."[13] It is not clear that Wilson ever directly consulted Rosenberg about her recommendation to adopt the new code, but, as Moreno speculates, it is reasonable to assume so given that she remained in the administrative milieu for several months following Wilson's debut.

On February 26, 1953, Wilson signed off on the new policy. Curiously, the memo he signed was tagged with a number: TS0118, standing for "Top Secret." The irony of this designation for a policy that was intended to make experimentation visible to its human subjects, while far from subtle, was telling of the atomic secrecy mindset. The "top secret" tag on the new policy would render it effectively useless. If only those with security clearance could ever see the code, then how would prospective human subjects benefit from it? The obvious outcome of this absurdity was that using uninformed human subjects was a practice that ran rampant over the next couple of decades. As Moreno so aptly puts it: "During that time federally sponsored human experiments seemed to career out of control in just about every direction."[14]

In the 1980s, the opaque cloak of secrecy covering the entire atomic enterprise began to lift, just slightly. While news reports of radioactive emergencies and "events" at Rocky Flats, Three Mile Island, and Church Rock, New Mexico, had forced the cloak to reveal some of its secrets, the public was still denied free access to information. In addition, the Cold War continued to suppress information in the name of national security and the arms race. President Reagan had instituted the largest military budgets in history, and the Russians were involved in similar activities that would insure, in both superpowers' minds, MAD—mutually assured destruction. The theory informing MAD was that if the two powers continued on the route of matching each other's nuclear arms stockpiles and methods for delivering their destructive powers, then we would stay in a "peaceful" stalemate. MAD concisely reveals both governments' atomic mindsets: MAD-ness.

In 1986, however, the cloak of secrecy tore. The U.S. House of Representatives Subcommittee on Energy Conservation and Power, chaired by Representative Edward J. Markey of Massachusetts, released a report (subsequently referred to as the Markey Report): *American Nuclear Guinea Pigs: Three Decades of Radiation Experiments on U.S. Citizens*. The Markey Report provided evidence of testing radioactive materials on humans through ingestion, injections, skin exposure, and a variety of other methods. The report's letter of transmittal to Michigan Representative John Dingell, the Chair of the House Committee on Energy and Commerce, states, "The report describes in detail 31 experiments during which at least 695 persons were exposed to radiation which provided little or no medical benefit to the subjects"[15] from the 1940s through the 1970s.

The report reveals that in many cases, human subjects—also known as "nuclear calibration devices"—did not grant their consent. "The government covered up the nature of the experiments and deceived families of victims as to what had transpired."[16] For example, between 1945 and 1947, 18 terminally ill patients, defined as having 10 or fewer years to live, were injected with plutonium "to measure the quantity retained by the human body."[17] These experiments were a collaboration of the Manhattan District Hospital at Oak Ridge, Tennessee, Strong Memorial Hospital in Rochester, New York, the University of Chicago, and the University of California at San Francisco. In the end, seven of these patients lived beyond the 10-year expectation and five lived longer than 20 years. But it was not until 1974 that any living patients or their families were informed about the plutonium injections. The Atomic Energy Commission, as a partial explanation for the secrecy, reminded investigators that even the word *plutonium* was classified at the time of the experiments.

Another instance involved elderly patients tested by the Massachusetts Institute of Technology from 1961 to 1965.

> Twenty subjects aged 63 to 83 were injected or fed radium or thorium to estimate internal doses and to measure passage of these doses through their bodies. Many of these subjects came from the nearby Age Center of New England, a research facility established to investigate the process of aging and the needs of the elderly. These experiments thus represent a perversion of the Center's original purpose, since feeding the subjects radium and thorium did not benefit them as individuals or the elderly population as a whole.[18]

In this case, the experiments merely benefited science, the scientists, and their related institutional apparatus. Once again, the scientists did know about the effects of radiation, but they claimed that they didn't know *enough*.

The Markey Report describes experiments with radiation given to pregnant women, to young children (as seen in the case of Simeon Shaw), and to residents of towns where "downwind" effects of fallout were studied by people who ate fallout-tainted crops and milk or were recipients of radioactivity in the air from atomic bomb tests. Frequently, the subjects had no knowledge of their status as human subjects. They were, as the report so boldly declared, human guinea pigs.

But if all of this information was under lock and key due to top-secret classifications, then how was it available to Markey and his associates, and, more importantly, when it was made available, why didn't the public seem to care? Here the story provides an even more telling aspect of the atomic mindset: indifference on the part of the national media, the public, politicians, and scientists—basically, nearly everyone.

In fact, two pieces were published in the popular media prior to Markey's 1986 report. *Science Trends*, a relatively small-circulation newsletter, published an article by editor Arthur Kranish in 1976. The article detailed the disturbing story of 18 people who had been injected with significant amounts of bomb-grade plutonium to illustrate how much toxin remained in a human body. Then, in September 1981, Howard L. Rosenberg wrote a cover story in *Mother Jones* telling of 89 cancer patients exposed to large doses of radiation between 1960 and 1974. These tests measured their "sensitivity" to the radioactive products. Many of them were tested without informed consent.

In both cases, the authors attempted to promote the stories in the news of the day. Kranish managed to get *Newsweek* and United Press International to run follow-up stories, but there was no national take on the stories beyond these brief announcements. Rosenberg's enjoyed a modicum of short-term media attention. He appeared on NBC's *Today Show* and the UPI picked it up, but "the attention only lasted for a day or two,"[19] he said. Further, Rep. Al Gore of Tennessee held hearings on the Oak Ridge experiments in September 1981, but the committee concluded, "the doctors performing the experiments could not be held responsible."[20] The topic of human experimentation with radioactive products drifted out of sight for another five years until the Markey report appeared.

Consistent with the other stories, the Markey report failed to attract public attention despite its attention-getting title. Few news articles discussed the report at the time of its release. Congress all but ignored the troubling findings. Such a tepid public reception might be explained by the timing of the report's release. The Cold War was still on the front burner of worldwide politics, and there was anxiety that the public might panic if they learned of what had been done to human subjects. Fear of panic can be traced to the military's observation that "the public in the United States had a 'hysterical and alarmist complex' about radiation that needed to be corrected to enable the U.S. to proceed with its testing activities," and suggested that "the process of correction would be a matter of reeducation over a long period of time."[21]

While the *New York Times* and *Los Angeles Times* did in fact run stories on the Markey Report, they were buried on back pages. "Not being on the front page, as a paper of record, it made it seem as if it were not significant,"[22] Markey's press secretary, Raoul Rosenberg, said. Markey, known as a critic of the administration's arms control policies and active in the nuclear weapons freeze moment, was not surprised by the lack of attention to the issues. "I was a voice in the wilderness in the mid-80s."[23]

One potent rationale for the media's disinterest in Markey's report is that the report did not name the subjects, or victims, of the testing. They didn't have a face, so to speak. Anonymity, common in scientific research, can be protective. It tends to reduce the messiness of human emotion that becomes so acute when the subject is personalized with a name: an actual human being who was someone's child or sibling or loved one. This practice of not naming would all change when Eileen Welsome, a dogged reporter for a small New Mexico newspaper, began digging.

In 1987, while searching for a good story, Welsome read an Air Force report detailing current cleanups of waste sites around the country. What caught her attention was that some of the sites had animals buried in them, animals that were radioactive. As Welsome writes in her Pulitzer Prize-winning book that would appear more than a decade later, "I have always loved animals and the disclosure caught my eye."[24] Her investigative reporter instincts immediately kicked into gear. "What kind of animals were buried in those dumps, I wondered, and why were they radioactive?"[25] Her curiosity led her to Kirtland Air Base, where she was given permission to scour technical reports about the dumps and the activities that had given rise to them.

The reports were "stiff with age and smelled of dust."[26] As you would expect of such reports, they were challenging to read, laced with technical terminology. One day she was about to pack it in when her eyes were drawn to a footnote that described a human experiment.

> The information jolted me deeply. One minute I was reading about dogs that had been injected with large amounts of plutonium and had subsequently developed radiation sickness and tumors. Suddenly there was this reference to a *human* [sic]

experiment. I wondered if the people had experienced the same agonizing deaths as the animals.[27]

Welcome wondered if she might be the first person to discover this horrific secret. "Naively I thought I might be the first to 'break this forty-year-old story.'"[28] But she learned soon enough that *Science Trends* and the U.S. Congress had beaten her to it. One piece was missing from the publications and reports, however: the names of the subjects. "They were known by code numbers only. I wondered who these people were . . ."[29] As she would discover during the course of a 10-year journey through countless government reports, medical records, and personal interviews, these human subjects were all part of a long and detailed history of intense, highly secret human subject experimentation. She peeled away the secrecy layer by layer and worked her way down to the actual names of many of the victims. Her first discovery was Elmer Allen, a man from Italy, Texas. Until she was able to cross-reference and verify his code and name, he remained CAL-3.

Welcome's uncovering and naming of the victims of radiation testing foreshadowed a most amazing pronouncement on December 7, 1993, by Secretary of Energy Hazel O'Leary. In front of the Washington press corps, O'Leary boldly revealed that America had been illegally and unethically testing thousands of human subjects with radioactive substances for 50 years. With relatively few words that day, O'Leary opened a Pandora's box of information: facts, figures, names, costs, horror stories. Aided by the Freedom of Information Act the next year, the American public and the world citizenry would come to know, through literally millions of pages of official documents, just how pervasive these practices had been.

The word *Nazi* and references to language associated with the Holocaust were becoming just too close, too real. O'Leary made the first such comment after her press conference in an interview with *Newsweek* magazine. "The only thing I could think of was Nazi Germany."[30] Ironically, her opening of the great radiation testing box would begin to reveal that, in fact, the Nazi connection had been made nearly 50 years before, as we saw in the memo written by Joseph Hamilton that alluded to the "Buchenwald touch." Meanwhile, although historians, through long-withheld documents and eyewitness testimonies, would start to tell us the stories of how much we knew about the risks of radiation, the testing would continue.

NOTES

1. R. Del Tredici, *At Work in the Fields*, caption to plate #5.

2. E. Welsome, *Plutonium Files*, 150.

3. Ibid., 152.

4. Ibid., 152.

5. Ibid., 155.

6. Ibid.

7. T. Szczygiel, *Schloendorff*, wings.buffalo.edu. A discussion of the case archived in UB Center for Clinical Ethics and Humanities in Healthcare.

8. Ibid.

9. See Document 3, point #2 in this chapter.

10. L. Musmeci-Kimball, *Conversation*, December, 2010.

11. K. Schneider, *The New York Times*, December 28, 1993. On the 1950 memo written by Hamilton to Dr. Shields Warren, a senior oficial of the AEC.

12. J. Moreno, *Undue Risk*, 168.

13. Ibid., 173.

14. Ibid., 188.

15. E. Markey, *Report* 1986.

16. Ibid., 1.

17. Ibid.

18. Ibid.

19. J. Braffman-Miller, "Human Experiments," 1995.

20. Ibid.

21. A. Makhijani, *Arms Control Today*, 2005.

22. J. Braffman-Miller, 1995.

23. Ibid.

24. E. Welsome, *Plutonium Files*, 2.

25. Ibid., 2.

26. Ibid., 3.

27. Ibid.

28. Ibid.

29. Ibid.

30. The Daily Beast, "The Dirty Little Secrets Of The Atomic Age," http://www.newsweek.com/id/125159 (accessed October 21, 2009).

CHAPTER 6

Dangerous Familiars Part 2: Three Cases of Human Testing, 1949–2011

The Jewish Holocaust is usually depicted as the darkest hour of eugenics. And indeed, it can certainly hold that dubious honor. The insidious practices of eugenics—a pseudoscientific and social movement aimed at improving the human race through controlled breeding—were not limited to the Nazis. While millions of Jews and other ethnic groups were targeted for eradication during World War II in order to create a superior Aryan race, eugenics continued after the war, inspired by the atomic mindset. In its hot pursuit of atomic knowledge, the atomic mind located "expendable" human bodies for scientific testing of radioactive products. Some of these were overt tests conducted through formal scientific methods in which unknowing individuals were subjected to radiation doses of various levels and then studied for the effects upon their minds and bodies. Other tests were also conducted scientifically, but instead of studying an individual or a small group of human subjects, these tests were perpetrated upon entire communities. And yet one other form of testing is not so overt and consciously conducted. These "tests" are often conducted after the fact when things have gone wrong, usually in medical treatments involving radiation therapies and technologies. In this chapter we will encounter three cases of human subjects testing: one of individuals, one of a large community 70 years ago, and one of medical patients in the present day.

THE "OATMEAL BOYS"
OF THE FERNALD SCHOOL

In the early 1900s, and then continuing well into the second half of the century, many states across the land operated schools for the "feeble-minded." Children who attended these schools were officially classified as "idiots or morons." Comparatively innocuous IQ tests and other analytical instruments were developed and used to measure mental potential in order to classify individuals. Some, those who scored low, were institutionalized in schools similar to Fernald— the oldest and one of the largest schools of its type in the United States—where some 2,500 residents, mostly children, were confined from the 1930s to the 1960s. As Walter E. Fernald, the founder of the Fernald School for Boys in Waltham, Massachusetts, firmly stated, "The social and economic burdens of feeble-mindedness are only too well known. The feeble-minded are a parasitic, predatory class, never capable of self-support or of managing their own affairs. . . . They cause unutterable sorrow at home and are a menace and danger to society."[1] Purportedly established to house and possibly rehabilitate these societal "menaces," Fernald was in fact a human guinea pig repository that would become a very convenient testing lab.

The conditions at Fernald and other such institutions were appalling. As one of the past residents described in 2004 to Bob Simon of *CBS News*, once the government defined you as unfit, "they could just dump you off in these human warehouses and just let you rot, you know. That's what they did. They let us rot."[2] In some cases, children were abandoned by parents who didn't know what else to do. Often, they were from very poor families, and the parents or guardians saw little recourse. One former resident of the school, Joe Almeida, recalls the day he was brought to Fernald at the age of eight. "I said, 'Wait a minute Dad. Where are you going?' He goes, 'Oh, you wait right there. I gotta go get the car.' And he went. That was the last I seen of him."[3]

But the "school" was not a school. There was little thought given to educating those who had been deemed uneducable. But they did work hard at manual labor. "They grew the vegetables they ate . . . sewed the soles on the shoes they wore . . . manufactured the brooms they used to sweep the floor," explains Michael D'Antonio, the author of *The State Boys Rebellion, The Inspiring True Story of American Eugenics and the Men Who Overcame It*.[4] But some of the boys did take part in science experiments, although these weren't educational in a true sense. Joe Almeida, for instance, "cut up the brains of severely retarded people who had died at Fernald. He cut them into thin slices so scientists could study them. Nothing ever came of the research, but the bits of brain are still there (in 2004)," reported Bob Simon.[5]

The most notable of the science experiments were those conducted under the aegis of "the Science Club." The club was the brainchild of researchers at the Massachusetts Institute of Technology, conveniently located within driving distance in Cambridge, Massachusetts. The scientists there, interested in the study of diseases and, yes, atomic radiation of uninformed subjects, saw an ample and available source of subjects who had been deemed something less than human. One of the leading scientists, Robley Evans, was a worldwide expert on radiation poisoning.

MIT scientists received consent for their experiments from the Atomic Energy Commission, who not only approved of giving trace radioactive elements to the boys but also gave permission to give larger doses to those who were suffering from gross physical and mental abnormalities, such as gargoylism and severe mental retardation. Further, in the event that a parent or guardian was known to the Science Club scientists, a letter was sent explaining that the club was an opportunity to "improve the nutrition of our children and to help in general [function] more efficiently than before." Radiation was never mentioned in the letters, of course, but it did state that the boys who took part received "many additional privileges. They get a quart of milk daily during that time, and are taken to a baseball game, the beach and to some outside dinners and they enjoy it greatly…"[6]

MIT's research partner for these activities was Quaker Oats Corporation. Quaker Oats was interested in how nutritional qualities of oats could be scientifically traced throughout the body and how to measure their effects. The Fernald School became a perfect Petri dish, perhaps only rivaled by the Bikini Atoll.

The boys of Fernald were exceedingly happy about joining the Science Club. They enjoyed many rewards just for taking part in the club. Taking trips to professional baseball games, the zoo, the beach, and other cultural attractions was something these boys had never done before. Some were even presented with a Mickey Mouse watch for Christmas.

Above all, the boys were given a daily treat. The food at the Fernald School was limited and not very tasty, so being able to have more food that tasted good was a big draw to members of the club. Between 1946 and 1953, 74 boys were given this special treat—oatmeal and lots of fresh whole milk on a daily basis. Unknown to them, the hearty breakfast contained trace radioactive elements, most notably radioactive iron produced by the MIT cyclotron. The scientists explained to the club members how important it was to finish their breakfast. "You had to drink the milk. That was the thing,"[7] said one of the club members, Gordon Shattuck, in an interview years later.

On a routine basis, scientists collected samples of the boys' urine and feces, and they periodically drew blood samples. These samples were taken away to MIT labs for analysis, but there is no known record that remains of the findings these experiments provided. It was top secret, after all.

What did remain were the potential and unwanted psychological and physical health effects wrought by the Fernald and MIT experiments on the boys. As with so many of the atomic guinea pig herd over the years, they did not have knowledge of what had been done to them until many years later. One of the now-grown boys, Fred Boyce, learned of his role in the experiments in early 1994 while listening to his car radio. He remembers being compelled to pull into a parking lot, shocked after hearing that *The Boston Globe* reported "a federal investigation had concluded that in the 1940s and '50s, scientists from MIT had conducted radiation experiments on unwitting children at the Fernald School "What! That can't be right. That's us! That's me!" declared a 57-year-old Boyce.[8]

Within three weeks of the *Globe* report, articles in the newspaper and other media sources had attracted so much attention that an esteemed group of politicians, scientists, and government officials set up a public hearing in the auditorium of the Fernald School. Led by Senator Edward Kennedy and Congressman Edward Markey, the meeting began with Kennedy's demand: "We want to know what was done in Massachusetts and in every other state where these experiments were conducted. We want to know what records exist, how great the dangers were, how much consent, if any, was obtained, how much harm was done." Kennedy was followed by a law professor from Boston University, George Annas, who stated, "The residents were used because they were convenient," and the abuse perpetrated on them was wrong "even by the standards of the time." Not surprisingly, a few remarks were made to downplay the dangers of the experiments. Bertrand Brill, a nuclear medicine specialist, speculated, "I don't think anything hazardous occurred." He also answered a question from the audience asking if he would have let his son take part in such experiments. Brill answered, "Knowing what I know now, I would have to say yes." David Lister, Dean of Research at MIT, chimed in by saying that the boys who ate the oatmeal and milk would have only a slightly higher risk of cancer.[9]

Both Fred Boyce and Joe Almeida watched the hearing on television and in person, respectively. Although they were not in the same space, they heard the hearings in the same way: as pronouncements by officials who, ultimately, would ignore what the situation called for: reparation for victims. They both believed that the focus on the harmfulness of radiation was not the issue at hand. Rather, the issue was justice, justice for having been held guinea pig prisoners of the state to do science's bidding. Like the Radium Girls, Bikinians, and Native Americans, they had been lied to, used, and then discarded. They thought that maybe they could get monetary compensation through the courts, but they also wanted apologies. In order to accomplish their goals, they knew they would have to take matters into their own hands.

This they did, with fervor and tenacity. They researched old files in the Fernald building and library. They made contact with other boys who had been used in the experiments. They found lawyers willing to help them. In short, they became researchers with a mission.

Their hard worked paid off, both in terms of monetary settlement and, eventually, some apologies. In 1998, Fred and Joe, along with 39 other former Fernald School residents, accepted a class action settlement with MIT and Quaker Oats for $1.85 million dollars. They also received apologies from President Clinton and the governor of Massachusetts. They did not, however, receive apologies from MIT and Quaker Oats. An apology could be interpreted as an admission of guilt and suggest further litigation and compensation.

Fred Boyce continues to pursue the case by locating and counseling former Oatmeal Boys. In 2002, Fred was diagnosed with colon cancer, received radiation, underwent surgery and chemotherapy, and was rid of the disease. Ironically, radiation finally played a healing role in his life, yet also could have been a factor in the genesis of his disease years ago when he was a member of the Science Club.

THE GREEN RUN AT HANFORD, WA: THIS IS A TEST, ONLY A TEST

Activities at the Hanford Nuclear Reservation in Washington State during World War II resembled a massive, well-organized storm.

Sudden and turbulent energy was harnessed to produce the plutonium and enriched uranium required to build atomic bombs. For three solid years, between 1942 and 1945, scientists, military personnel, and workers labored furiously around the clock. After the bombs fell on Japan, the storm settled—momentarily. Then a second storm loomed, this one seeded in Eastern Europe. In September of 1949, the Russians exploded their first atomic bomb. Anxiety spawned by this event radiated out to politicians, scientists, and especially the American public. Nobody, though, was more anxious than the military. Enriched uranium and plutonium production would need to increase, and this need would eventually become the seed of yet another massive energy storm located again on site at Hanford. A bit of Hanford Reservation's birth and how it became a gargantuan enterprise provides the background for putting the Green Run into perspective. One character played the lead role.

General Leslie R. Groves (the man who wrote the letter introducing *Dagwood Splits the Atom*) was hell bent to deliver plutonium to Los Alamos for the first atomic bomb. He was legendary for his ability to get things done. Everything from the startup of the Manhattan Project in Chicago to the on-the-ground hunt for all available uranium reserves in the world had been in large part orchestrated by Groves. Building one of the largest industrial facilities under an impenetrable cloak of secrecy was just one more monumental task to complete.

In early 1942, Groves wanted to build a massive plutonium complex in either the Argonne Forest outside Chicago (and thus in proximity to the University of Chicago, where many of the physicists involved in the atomic bomb project resided) or at Oak Ridge, Tennessee, where the then-largest industrial complex in the world was already underway. Particular problems associated with large-scale plutonium production quickly ruled out these two sites. Argonne was just too close to a large urban area, and the necessary acreage was not available; Oak Ridge was already gobbling up massive infrastructural resources (especially electricity), and the site was simply too restricted for such a massive effort.

Just how massive was the proposed project? Groves laid out necessary requirements for the plutonium production venture that included: (1) 25,000 gallons of water per minute, (2) 100,000 kw of electrical power, (3) a hazardous manufacturing area no less than 190 square miles, (4) laboratories and local inhabitants no closer than 10 to 20 miles from the facility, (5) no highways or railroads closer than 10 miles, and (6) a climate that "should have no effect on the process."

This final criterion would elude even Groves's impressive talent for control.[10]

Once the construction of the Hanford facility began in 1943, its enormity became obvious. Three massive separation plants, dubbed "Queen Marys" by the Du Pont workers, were constructed. These Queen Marys dwarfed the huge ocean liner that was their namesake. Each building was 800 feet long, 65 feet wide, and 80 feet tall, and all were sealed with 35-ton covers placed by cranes. Each Queen Mary consisted of concrete containment cells with seven-foot-thick walls and tops that were six feet thick. These massive containers were built to isolate the highly radioactive products resulting from a process that involved dissolving radioactive uranium slugs in concentrated nitric acid. The three main plants were divided into 40 compartments in which this process took place. As Richard Rhodes describes it in *The Making of the Atomic Bomb*:

> The slugs then would move in shielded casks on special railroad cars to one of the Queen Marys, where they would be dissolved in hot nitric acid . . . The liquid solution that the slugs had become would move through these units by steam-jet siphoning . . . The end product would be radioactive wastes, stored on site in underground tanks, and small quantities of highly purified plutonium nitrate. "When the Queen Marys began to function," the physicist Leona Marshall remembers, "dissolving the irradiated slugs in nitric acid, great plumes of brown fumes blossomed above the concrete canyons, climbed thousands of feet into the air, and drifted sideways as they cooled, blown by the winds aloft."[11]

Uranium slug processing at Hanford was an intense and complex endeavor, producing a scant few pounds of precious material—enough enriched uranium (U-235) and plutonium (U-239) to feed the two bombs that were dropped on Japan in August 1945. The millions of gallons of water, massive amounts of electricity, the total development effort of the Hanford, Los Alamos, and Oak Ridge facilities, and the thousands of workers and scientists employed there gave to humankind enough precious, potentially explosive, and tremendously toxic materials needed for the war effort. With barely time to take a breath, however, the process of creating weapons-grade plutonium would soon demand even more: more efficiently produced stockpiles of the precious radioactive element as a second storm loomed on the Eastern horizon.

In September 1949, the Russians exploded their first atomic bombs, and the political and military storms that this created would rival the storms at the end of World War II. The processes for creating more plutonium at Hanford had been improved to some extent, but there was still not enough of it being created to keep up with the perceived pace of Russian bomb development. More plutonium was needed, and there was no one better to make that happen than General Groves.

To meet this demand for more bomb-grade plutonium, Groves was pushing Hanford hard: "Get that stuff up to Los Alamos as quick as you can"—the cooling time maybe now is a hundred days—"Let's get it down to fifty, maybe even forty,"[12] remembered pre-eminent health physicist Karl Morgan in a 1995 interview about what it was like working for Groves at Hanford. Morgan explained that "If they reduced the cooling time of the uranium slugs before they were dissolved in acid, the plutonium could be delivered earlier to Los Alamos (and please General Groves) but much more radioiodine would escape into the Hanford environment."[13] Thus, the purpose of reducing the cooling time would allow for more plutonium to be delivered more quickly to the bomb makers, but the uranium product would be "green" because it hadn't ripened yet. The product, when it was cooled in its green state, would release a significantly more potent mixture of radioactive byproducts, primarily iodine-131, into the atmosphere. Here was the paradox, one that the scientists and military officials understood well: shorten the cooling period and create more plutonium with more radioactive emissions or reduce the emissions by lengthening the cooling period.

This was the price to pay for making more plutonium quickly. Green processing it would be—even though the health and environmental risks were certainly higher with the shorter cooling period. "Haste makes waste" was literally true in this case, and the waste was dangerous stuff. However, this is really nothing new under the sun for the atomic mindset. They knew that iodine-131 could cause considerable thyroid damage, but the Russians were building A-bombs. Groves knew exactly what to do: cut down that cooling time so the United States wouldn't lose face. A few thyroid glands were a small price to pay.

On the night of December 2, 1949, the decision was made to release the emissions from the green uranium solution that had only cooled for 16 days, many fewer days than even the low end of what experts like Karl Morgan had imagined. On top of the decision to only wait 16 days was the issue of the *actual* weather that surrounded this atomic

storm. As it turned out, the weather on the chosen night was unpre-
dictable: not a good thing for dispersing radioactive emissions. If
winds are low, then the somewhat heavy gasses can stay confined in a
smaller area and thus drop their byproducts in a more concentrated
fashion. They had hoped for higher winds to dilute the gases by
spreading them more widely. Even though some meteorologists working on the Green Run experiment cautioned against release, it happened. In the end, a Department of Defense meteorologist made the
final decision to go ahead.

On the evening of the test, weather conditions were even worse than
anticipated, with rain and snow carrying heavier-than-anticipated
deposits of stack gases northwest and southwest of the stacks. As Carl
C. Gamersfelder, the number two man in Hanford's Health Instruments Division, recalled, "The clouds wandered off and went down
to the Columbia River Valley, turned around and came back and wandered off to the east."[14] As it wandered, the radioactive cloud would
spew 7,780 curies of iodine-131 and 4,750 curies of xenon-133,
another highly volatile radioactive element, over 75,000 acres of
Washington and Oregon, an area occupied by some 2 million inhabitants, not to mention countless domestic livestock and wildlife.

The whole incident would remain totally secret until 1986, and even
then (and up to the present day), much information about the Green
Run remains undisclosed or has been conveniently lost. However, in
response to the Hanford Education League[15] (a citizens' group formed
to provide information about the Hanford enterprise), the truth—or
some of it—was told through documents released by the Freedom of
Information Act. Their investigations not only told the story of the
Green Run in some detail but also included a surprise: the Green
Run, apart from being a hastened plutonium-producing exercise, also
was used to test Air Force monitoring equipment. For years following
the 1949 spewing of radioactive gases, scientists gathered samples of
blood from local residents, tested milk and agricultural products for
radiation levels, and interviewed residents about their health. Just like
the workers in the Manhattan Project, the Native American miners,
the Bikini Islanders, and so many others, they were never told why they
were monitored. "This is a test," the Civil Defense Department was
wont to say, "and only a test."

And it wasn't the only test. The tests would continue at Hanford
until 1972, and although not as large as the Green Run, the cumulative
amounts of radioactive debris would rival many other large human subject experiments during that time. Haste, fear, secrecy, and curiosity

pushed the atomic mindset beyond the bounds of reason. The result, as Karl Morgan said, was that too often, [people like] Groves won in this contest, and only gravestones testified to the unwitting losers in the silent struggle.[16]

RADIATING THE TWENTY-FIRST CENTURY: X-RAYS, CT SCANS, AND SECONDHAND RADIATION

Scott Jerome-Parks was known as the proverbial man who would give you the shirt off his back. He worked in New York City as a computer and systems analyst for CIBC World Markets, a job he had just taken in September 2001. He worked not too far from the World Trade Center, so after two passenger jets flew suicide missions into the immense towers on September 11th, Scott volunteered over the coming weeks to help in any way he could. He "donated blood, helped a family search for a missing relative, and volunteered at the Red Cross, driving search-and-rescue workers from what became known as 'the pile,' " reported *NY Times* correspondent Walt Bogdanich in a January 2010 story titled "Radiation Offers New Cures, and Ways to Do Harm."[17] Four years after the 9/11 tragedy, Scott contracted what he thought was a sinus infection that would just not go away. He tried treating himself but eventually went to a doctor. In December 2004, he learned that he did not have an infection, but instead had a cancer of the tongue. For a nonsmoker and only occasional drinker of alcoholic beverages, it was quite a shock. But, being a careful thinker and man of considerable patience, he researched his alternatives in as much detail as you can under circumstances in which time is of the essence.

His investigation of various options led him to choose St. Vincent's Hospital—also the primary treatment center for 9/11 victims—as the place to receive radiation therapy on the base of his tongue. The hospital was quickly becoming an institution that was seeking new ways to improve its financial situation in the highly competitive environment of medical treatments. To move into this growing market, St. Vincent's had invested in a new linear accelerator technology, developed by Varian Medical Systems and managed through software designed by Aptium Oncology, that touted "smart-beam technology," technically known as intensity modulated radiation therapy or just IMRT. In *Principles of Information Systems*, authors Ralph M. Stair, George Reynolds, and George W. Reynolds explain that, "the SmartBeam IMRT combines an x-ray and *radiation technology* [sic] into one device that rotates around the patient delivering radiation at precise intensities from any angle."[18] The heart of the IMRT's hardware is the multileaf collimator that consists of more than 100 metal leaves controlled by software. The leaves are programmed to open and close, thus shaping and directing the radioactive beam toward its intended cancerous target. The collimator and its attendant software would become critical players in Scott Jerome-Parks' tongue treatments.

As Aptium's website explains, the company aims

> To deliver market-leading oncology solutions that help hospitals and physician practices optimize cancer care, enabling patients to achieve longer, better lives … If a company is judged by the company it keeps, our clients say it all. From academic medical centers to community hospitals across the country, our clients rely on us for best-in-class consulting and management services that drive business success. Your cancer program will benefit from the combined brain trust, resources and best practices our experience has provided us over the years.[19]

In Scott Jerome-Parks' case, the best practices included fatal flaws involving Varian, Aptium, St. Vincent's Hospital and, most tragically, Scott Jerome-Parks.

Scott underwent four radiation treatments over several months, and they all went according to plan. In March 2005, he entered the hospital for a fifth session. The previous four treatments had caused some bone weakening in the jaw area, so the doctor, Anthony M. Berson, had a medical physicist, Nina Kalach, check the calibrations on the SmartBeam IMRT. She recalibrated the machine's software but in doing so encountered some error messages, and the computer froze. A final prompt asked if Ms. Kalach wanted to save the changes she had made. She said yes, with Dr. Berson's approval, to the software.

Scott was alone in the sealed treatment room when the irradiation started. What nobody realized in the moment was that the computer had frozen again. Not perceiving any problems, the medical staff conducted a second round of treatment the following day. After the treatments concluded on the second day, a friend of Scott's, Paul Bibbo, paid a visit to see how things were going. The scene he entered was shocking. Scott's head and neck were so swollen that he described them as blown up. Scott was writhing in pain, tossing from side to side. His distraught wife, Carmen Jerome-Parks, didn't know what to do to help her husband. She stood at the foot of the bed in the darkened room and prayed.

On the next day, the hospital provided a psychiatrist for Carmen. After Carmen's session with the psychiatrist, the hospital administered Scott's final radiation treatment. Following this final treatment, a test was run on the IMRT system and it found that, indeed, a grave error had occurred. The leaves of the collimator had been wide open! Scott's neck and base of his skull all the way to the larynx had been exposed. A later investigation revealed that he had received seven times the prescribed dose.[20]

For nearly two years, Scott struggled with the incurable ailment that would have him take constant pain medications and gradually lose his eyesight, hearing, and balance. In February of 2007, Scott finally succumbed. One close friend of the Jerome-Parkses, Ms. Giuliano, said, "His [Scott's] hope was that 'My death will not be for nothing.' He didn't say it that way, because that would take too much ego, and Scott didn't have that kind of ego, but I think that it would be really important to him that he didn't die for nothing."[21] Scott's wish sounds poignantly reminiscent of Katherine Schaub's case 70 years before. One could say that, like Katherine, Scott died for science.

Many stories like Scott's exist and could be told. Several books could be written on just these kinds of horrible incidents; in fact, some already have been. Most of the time, stories of radiation treatment error focus on faulty human operating procedures and/or the breakdown of the machinery itself. However, at this point I want to change the story somewhat by making the case that such occurrences operate within the context of human subject research. That is, radiation is regularly used to investigate, diagnose, and treat cancer, broken bones, tooth problems, internal organ disease, and more. Yet some of these uses of radiation are underregulated, or there is simply no consistent, centralized regulation. "In 17 states, the technicians or therapists who operate the radiotherapy machines do not have to be licensed. And eight states allow therapists to perform medical imaging other than mammography with no credentials or educational requirements."[22]

In cases where more scrupulous regulation and oversight exist, the data used to investigate the numbers of accidents is sometimes inadequate because such reporting is not required. "With no single agency overseeing medical radiation, there is no central clearing house of cases. [Thus], regulators and researchers *can only guess* [emphasis mine] how often radiation accidents occur."[23] When the information does reveal itself, the facts are very disturbing: a Philadelphia hospital reported in 2010 that it had covered over the fact that 90 patients had received the wrong radiation dose for prostate cancer treatment; a Florida hospital "disclosed that 77 brain cancer patients had received 50 percent more radiation than prescribed" using a linear accelerator similar to the device used on Scott Jerome-Parks; "in 2009, the nation's largest wound-care company treated 3,000 radiation injuries."[24] Some of these victims of radiation mistakes have received settlements after going through litigation. However, the government is known to levy pitifully small fines on the institutions that perpetrated the overradiation. In the case of Scott Jerome-Parks, for example, St. Vincent's hospital was fined $1,000, and that was for failure to catch the error in the software, not for overdosing Scott.

Secrecy abounds in the atomic mindset. Even though the issue of medical radiation usage has been a well-documented problem for more than 30 years, the information is often held back or simply not kept.[25] If, for instance, you want to find out where mistakes have happened or who was involved, state laws can prohibit the disclosure of such information in the hopes that it will encourage the reportage by hospitals. There is little evidence that this strategy of encouraging reporting of incidents works very well. "My suspicion is that maybe half of the accidents we don't know about," said Dr. Fred A. Mettler, Jr., an author of several books dealing with radiation in the medical world.[26]

Problems with radiation dosages are now receiving some attention, to positive effect. Thanks to the investigative work of people like Harvey Wasserman, Walt Bogdanich, Eileen Welsome, Edward Markey, and others over the past 30 years, the public and government officials are attending to the problems of radiation in medicine and research. Echoing the antismoking campaign's thrust against second-hand smoke, awareness of the dangers of second-hand radiation is on the rise. Cancer patients sent home after treatment with radioactive iodine have contaminated hotel rooms and set off alarms on public transportation, a congressional investigation found. They've come into close contact with vulnerable people, including children and pregnant women, and the household trash from their homes has triggered radiation detectors at landfills. Not surprisingly, Rep. Markey has been pulled into this most recent radioactive fray. Markey explained in a letter from his Congressional office in October 2010 that the problems arose from a relaxing of guidelines by the Nuclear Regulatory Commission years ago. The old guidelines had required that patients receiving radioactive iodine treatments to shrink tumors stay in the hospital a few days. The relaxed guidelines allow patients to be released immediately upon completion of the treatment. "There is a strong likelihood that members of the public have been unwittingly exposed to radiation from patients. This has occurred because of weak NRC regulations, ineffective oversight of those who administer these medical treatments, and the absence of clear guidance to patients and to physicians," Markey wrote.[27]

An Associated Press article reporting on Markey's announcement ends, "It's unclear whether the radiation exposure occurs at levels high enough to cause harm." [28] As always, we don't know enough yet. We will just have to go on testing.

NOTES

The title for this vignette was borrowed from Frances E. Dolan's *Dangerous Familiars: Representations of Domestic Crime in England, 1550–1700*, a thorough study, using literary and legal documents from Renaissance England, of the often horrific crimes committed by women upon their "familiars": husbands, children, siblings, and servants, among others whom they knew intimately.

1. E. Welsome, *Plutonium Files*, 232.
2. B. Simon, *60 Minutes*, 2004.
3. Ibid.
4. Ibid.
5. Ibid.
6. E. Welsome, *Plutonium Files*, 235.
7. Ibid., 231.
8. Higgins, *The Boston Globe*, January 18, 1998.
9. M. D'Antonio, *State Boys*, 247–248.
10. L. Groves, *Now*, 71.
11. R. Rhodes, *Making*, 604.
12. R. Del Tredici, *At Work* (interview with Morgan), 132–134.
13. Ibid.
14. J. Staggers, *Love the Bomb*, 34.
15. Hanford Education League 2010, http://www.hanfordchallenge.org/hanfords-history/the-hanford-x-files/.
16. Office of Health, Safety and Security, www.hss.energy.gov/healthsafety/ ohre/roadmap/histories/0475/0475c.html.

17. W. Bogdanich, *New York Times*, January 23, 24, 2010.

18. R. Stair and G. Reynolds, *Information Systems*, 151.

19. Aptium Oncology, 2011, www.aptiumoncology.com.

20. W. Bogdanich, *New York Times*, January 23, 24, 2010.

21. Ibid.

22. W. Bogdanich, *New York Times*, February 5, 2010.

23. Ibid.

24. Ibid.

25. H. Wasserman and N. Solomon, *Killing*, 125–139.

26. W. Bogdanich, *New York Times*, January 23, 24, 2010.

27. E. Markey, October 2010 letter to NRC concerning early release of patients treated with radioiodine (distributed by AP).

28. Associated Press, *Daily Mining Gazette*, October 10, 2010, 5A.

CHAPTER 7

Doing It in the Backyard: Alba Craft, Inc., Part 1

In the spring of 1993, Daryl Kimball, a young research associate for the Washington, DC-based Physicians for Social Responsibility (PSR), was delving into a plethora of documents recently released through President Clinton's Openness Initiative concerning Cold War activities. Kimball considered the work very significant, although numbing due to the sheer volume accumulated and kept under wraps for more than 40 years. His task was to scour these documents for information concerning formerly utilized atomic weapons production facilities from the 1950s. They had recently been released by President Clinton's directive overturning President Reagan's directive 10 years earlier that sealed files deemed dangerous to national security. PSR wanted to gather more information about known facilities for the risks they posed to public health, but just as importantly, they wanted to discover clandestine facilities that had fallen off the radar screen during the Cold War period. Kimball knew that these newly declassified documents would reveal important information pertaining to nuclear waste and related health hazards. However, he didn't know how close to home one of these discoveries would take him.

One of the documents was a map of sites across the country at which former atomic production sites had been identified. Daryl was familiar with the map, but this was an updated version that had a new entry: Oxford, Ohio. He immediately called a contact at the Department of Energy to request more information about this new spot on the map. He was provided an account for October 1952 to February 1957, when Alba Craft Laboratories, Inc., was under contract to National Lead of Ohio (NLO)—a primary contractor to the

U.S. Atomic Energy Commission (AEC). NLO developed and machine-threaded natural uranium metal to be used at the AEC Savannah River, South Carolina, Site. NLO also performed hollow drilling and turning of uranium metal to be used in the Savannah River and Hanford, Washington, nuclear reactors. NLO machined several hundred tons of uranium at the Alba Craft site.[1]

Then came the kicker. The address for the Alba Craft site was in Kimball's hometown at 10-14 West Rose Avenue, Oxford, Ohio. He had grown up blocks from this address and gone to school and college there. He knew exactly where the facility was located—right in the middle of a residential district near the ball field and playground of his old school, Stewart Junior High. The building sat in a neighborhood surrounded by homes of long-time Oxford residents. He even vaguely recognized the name of the facility—Alba—although the spelling was different than that of a professor of industrial arts at the local university, Bill Albaugh, who had lived on Main Street in the house right across the street from the site.

Kimball could not recall hearing of any such facility in Oxford, which was odd given his background. His father was a professor of history at Miami University who taught and wrote on nuclear issues. His mother, Linda Musmeci Kimball, a founding member and president of the Oxford Citizens for Peace and Justice (OCPJ), was deeply committed to public issues related to nuclear proliferation. But Daryl had never heard that uranium processing had been going on in their neighbors' backyards.

His startling discovery called for some immediate action and led him to telephone his mother. Her role as director of the OCPJ Peace Center made her the perfect recipient of such news. She was keenly aware of the intricacies of exploring sensitive issues concerning public health and radioactive contamination. His mother, it turned out, was aware of no such activity or facility in the Oxford area. There had been much news concerning nuclear waste and contamination in nearby Fernald, Ohio, and the Mound facility in Miamisburg, Ohio—two of the large production facilities that had fed enriched uranium to the atomic weapons industry for several decades—but she knew of no site located in small-town America, much less Oxford. Linda was perplexed. After hanging up with her son, she went over to West Rose Avenue to take a look at the building site.

When she arrived at the site, she found a nondescript, single-story, L-shaped concrete-block building. There were no signs on the building, but she could hear some faint sounds of music coming from what seemed to be a radio inside. The building itself was sound but badly needed painting and repairs to the concrete surfaces of the exterior walls. It was surrounded on two sides by houses and on a third by an apartment complex serving students of Miami University. To the west there were no houses, just a field. The field stretched away from the railroad tracks, across a field and playground area, and stopped at the walls of the junior high school. The railroad tracks, separating the field from the site, played an important role in the history of the facility, as Linda would soon discover. There was nothing she could do at that moment, but there would be—and very soon. So Linda went home to prepare. She would bring attention to what certainly appeared to be radioactive property within the city limits and near a school. She embarked on a project of discovery and exposure that would involve the Department of Energy, the U.S. Senate, the city government of Oxford, its citizens, and many other players—and would span more than three years.

The rest of the post–1993 Alba Craft story will be told in a subsequent chapter. Now, however, we return to 1952 to see what occurred at this mysterious building and who were the people who worked in it and lived so close by in this midwestern college town.

THE URANIUM MILLING COTTAGE INDUSTRY OF 1950S AMERICA

Much ink has been spilled in the past few decades describing the enormous projects that created the U.S. nuclear buildup. The key sites that were developed during and immediately after World War II included processing and production centers at Fernald, Ohio; Hanford, Washington; Oak Ridge, Tennessee; Rocky Flats near Denver, Colorado; Pantex outside Amarillo, Texas; Los Alamos Labs, New Mexico; and the Savannah River site in South Carolina. Other major sites dotted the growing nuclear network across the United States. These sites composed the backbone of the enormous atomic development project that was completely under the auspices of the U.S. military and the AEC. They employed, in total, well over a hundred thousand workers and were the places that uranium was processed and turned into other materials, such as plutonium for the making of warheads. These were hungry places; they needed to be fed constant supplies of low- and highly enriched uranium to keep the atomic beast alive.

Consequently, the AEC and the military developed a vast array of what could be called "uranium granaries" that constituted a cottage industry for a number of entrepreneurs. Under highly secret contracts, scores of uranium processing facilities were spawned to mill uranium into usable forms for atomic energy and weapons development. These facilities cropped up in old warehouses, abandoned manufacturing sites, just about any place where processing could be done. The products of this cottage industry were sometimes manufactured crudely. Workers would grind uranium on lathes and other relatively small milling machines. These uranium slugs, as they are called, were lathed and drilled until they met certain specifications of quality and size. They were stored at the facilities for short periods of time in containers such as barrels or even feed bags (anything sufficient to conceal the true identity of the contents). The barrels or bags of slugs would eventually be loaded into trucks or railroad cars to be transported to one of the great uranium processing facilities down the line.

Secrecy surrounding these smaller sites was intensely maintained. The need for secrecy in the atomic mindset gripped the general public and government officials alike. But how could facilities reside *undetected* near and within busy residential areas, occupy known building sites, and employ workers? How is it that no residents knew of their existence? Were railroad workers and others involved in transporting the "goods" also unaware? Did local governments not attempt to collect taxes on these properties? Wasn't it obvious that something profitable was going on? All of these questions cannot be answered here, but one case of how such highly secret events occurred right under the noses of a citizenry is that of Oxford, Ohio, and Alba Craft, Inc.

OXFORD, OHIO AND THE GENESIS OF ALBA CRAFT, INC.

Oxford, Ohio, is located in the rolling hills of the Miami Valley in the southwest corner of Ohio, about 40 miles west of Dayton and 30 miles north of Cincinnati, and is the home of Miami University. Oxford is a quintessential midwestern college town. Founded in the early 1800s, it was originally deeded through early government land development as a classic "mile square" with the boundary of the town consisting of one mile on each side of a square of land. Residents of Oxford aptly refer to the bucolic heart of the now much larger city as "The Mile Square": not at all somewhere you might expect a uranium-processing facility that prepared fissionable materials for atomic warheads. But as one professor of industrial arts at Miami University saw it, the Mile Square of Oxford, Ohio, was a perfect setting for just such an endeavor.

Bill Albaugh was a bright and enterprising sort from a very early age. Born on March 3, 1901, in West Milton, Ohio, he came to Miami University in 1924 to study in the newly formed industrial arts program. After only two years, as a sophomore, he was appointed as a part-time instructor in the program. He received his bachelor's degree in 1928 and was hired full-time as an instructor while pursuing a master's degree that he earned in 1933. Upon receiving the advanced degree, he assumed a position as a professor of industrial arts, a position he held until 1952, when he left the university to own and operate Alba Craft, Inc.

During his time at Miami, he tinkered with a number of sundry inventions. He developed some unique fishing lures, created an unusual cigarette filter, then became very interested in redesigning industrial machines, particularly lathes and related machines for milling and grinding raw materials. In addition, he learned how to fly airplanes and soon became a flight instructor. This new qualification led to an involvement with the Civil Aeronautics Administration during World War II as a liaison officer, and eventually he directed the Civil Pilot Training Program in Oxford. After the war, he served as director of veterans' affairs at the university until 1952.

Albaugh's experiences and connections fostered strong relationships with those technologies required to mill uranium and ultimately ushered Albaugh into the highly secret world of atomic weapons development. He had the necessary security clearances from the Atomic Energy Commission. He knew how to communicate through military channels. He was trusted by the various entities involved. He was a machine shop technician and teacher. He had all the qualities one needed to run a startup uranium milling cottage industry— an industry that, along with the uranium prospecting boom in the Southwest United States, would draw upon the entrepreneurial spirit and sprout hundreds of atomic cottage industries across the country.

THE ATOMIC ENERGY COMMISSION GIVES THE GREEN LIGHT

The quest for uranium increased to a frenetic pace by the 1950s. Raw uranium ore was now being prospected and mined as increasing amounts of potential fissionable materials were needed. Overseeing this project of atomic development and its entire infrastructure was the AEC. This five-member, civilian-led government commission formed by President Truman in 1946 determined atomic weapons development, as well as the United States' fledgling atomic energy needs. The commission managed four subcommittees that included military, civilian, and government officials. Their charge was clear cut: beat the Russians to bigger and more powerful bombs.

By the end of World War II, total costs for bomb development had topped $2 billion, but this became a relative drop in the bucket over the next decade. By 1952, infrastructure expense alone required an additional $2 billion, and operating costs had more than doubled. The number of employees to support this burgeoning business had tripled.[2] By 1953, Savannah River, a production facility consisting of immense buildings necessary for housing massive machines, had become the largest construction job in U.S. history, overtaking the old record set by Oak Ridge during the war. New plants built in Paducah, Kentucky; Portsmouth, Ohio; Fernald, Ohio; St. Louis, Missouri; Livermore, California; Denver, Colorado; and other places added to the gargantuan enterprise. In addition, Oak Ridge, Tennessee; Hanford, Washington; and Sandia Laboratories in Albuquerque, New Mexico, had at least doubled in size from their wartime beginnings.

Then, on October 31, 1952, President Truman once again decided to explode a bomb. This time, however, the bomb was a hydrogen bomb. The test took place in the Marshall Islands of the Pacific Ocean and was more powerful than any of its predecessors: The explosion equaled 10 million tons of TNT. The explosion obliterated one of the islands. Not to be outdone, and in just a few short months, the Russians countered by exploding an even more powerful device. The Cold War had begun, and with this new race to the top, the AEC needed more and more uranium prepared for the maw of the atomic leviathan. Enter the uranium-processing cottage industry and Alba Craft, Inc.

On the June 27, 1951, Eugene "Bill" Albaugh signed a contract with National Lead of Ohio (NLO) to begin operation of Alba Craft, Inc. The general terms of the contract were straightforward, as straightforward as any government contract might be. The "Scope of Work" in Article II stipulated that the Subcontractor (Albaugh) would "promptly perform ... (a) the labor, supervision, equipment, and facilities necessary to perform machining operations on materials supplied by the Contractor (NLO), and (b) machining operations in accordance with the specifications of the Contractor."[3] The contract also stipulated that "the Contractor on each of the Subcontractor's working days shall supply to the Subcontractor a minimum of one hundred and fifty units of the materials in Section II (a) ... (and that) the number of working days may be increased to include Saturdays and Sundays."[4]

Considering its legalistic and objective tone, critical urgency is obvious in the language of the contract, shown most dramatically in the all-uppercase words "CONFIDENTIAL SECURITY INFOR-MATION" stamped on the cover page.

TECH.-9 **CONFIDENTIAL** NATIONAL LEAD COMPANY OF OHIO   June 24, 1953
SECURITY INFORMATION Contract No. AT(30-1)-1156

PLANT ASSISTANCE AND DEVELOPMENT PROJECT PROPOSAL AND AUTHORIZATION

1. Project Number _____ 41-1
2. Title ____ Turret Lathe Machining of SRO Thread
3. Objectives To develop a machining operation, using Morey turret lathes for fabricating a finished SRO slug.

4. Background Preliminary test work at Alba Craft Inc. indicates an opera-tion on a #4 Morey turret lathe consisting of a small form tool and lead screw to machine a SRO slug complete in one loading operation. The work piece can be supported by using a stationary chuck mounted on the hexagon turret of the lathe. A roughing tool can be mounted ahead of the form tool, with double radius and cut-off tools mounted on the rear cross slide so as to enable one cut across the piece with one subsequent cut to radius and cut-off. Sequence of operations as follows: (See Attached Program).

5. Methods and Scale Contemplated 2,000 slugs are to be fabricated on a sub-contract operation by Alba Craft Inc., Oxford, Ohio. These will be returned to FMPC for inspection.

6. Estimated Manpower Required (Man Months) A. Technical __1__ B. Non-Technical____
7. Starting Date ___June 1, 1953_____ Estimated Completion Date August 1, 1953
8. Estimated Total Cost _$5,060.00_

   (A) Salaries and Wages ____$500.00____   (E) Indirect Costs ____$2500.00__
   (B) Analytical _____   (F) Overhead _____$ 400.00__
   (C) Materials and Operating Supplies $1000. (G) Contingency _____$ 660.00__
   (D) Additional Equipment Required _____

9. Estimated Savings to be Effected _Possibly $0.20 per slug_
10. Method of Reporting __Weekly Status and Summary Technical Reports._

11. Budget Activity No. __(14B) 2603_

   Signed   *Robert Mui*
   Approved   *Imleborsi*
               Department Head
               T. L. Cutthd
               Division Director

               _____
               Plant Manager

   Disapproved _____
   Reason _____

   RESTRICTED DATA
   AEC Branch Chief
   This document contains restricted data as defined in the Atomic Energy Act of 1946. Its transmittal or the disclosure of its contents in any manner to an unauthorized person is prohibited.

   Distribution: **CONFIDENTIAL**
                 SECURITY INFORMATION

[left margin handwritten:] TECH.-90 CHASS FILE Classification Cancelled Or Changed To UNCLAS By Authority Of PATTON Date 9/9/93 By 326-734-2047 Box : 38 FOUPOR : Process Development 4MN 326 R301 FPCCP41

Example of a 1953 top secret uranium milling operation contract for Alba Craft, Inc. (Provided by the Oxford Citizens for Peace and Justice archive.)

CONFIDENTIAL NATIONAL LEAD COMPANY
OF OHIO
SECURITY INFORMATION BOX 158, MT. HEALTHY STATION
CINCINNATI 31, OHIO

May 5, 1953

SUBJECT   PROPOSED PROGRAM TO DEVELOP A TURRET LATHE OPERATION FOR MACHINING
          THE SRO SLUG

TO   J. M. Ciborski

FROM   J. Farr   UNCLASSIFIED
                                                     DOE/SA-20
REFERENCE   06-01-93
                                                  D. A. Hughes

A program will be initiated to develop a machining operation on a
turret lathe for fabricating the SRO slug. This operation will
consist of rough turning, threading, radiusing and cutting off.
A more detailed explanation of the operation is as follows:

1. Load uranium rod in feed tube of turret lathe with approximately two inches protruding from the collet in the spindle.
   Close collet.

2. Move cross slide towards spindle against stop and move
   carriage towards spindle against stop, thus machining the
   two inches of protruded rod.

   NOTE: This enables the rough rod to be held by the sta-
          tionary chuck mounted on the hexagon turret.

3. Move turret on ram towards spindle against stop and lock in
   place. Open spindle collet and feed the two inches of already machined stock through the stationary chuck mounted
   on the hexagon turret. Close stationary chuck on machined
   surface of rod.

4. Close spindle collet.

5. Move carriage away from spindle against stop.

6. Move cross slide towards work piece against stop.

7. Engage feed lever on carriage for .020 inch per revolution
   thus roughing the work piece and machining the formed thread.

8. Back-off cross slide to stop, thus causing the double radius
   tool to machine a double radius on the work piece and the cut-
   off to cut off the previous two inches of machined surface.

9. Move cross slide towards spindle in order to remove the double
   radius and cut-off tools.

10. Open spindle collet and stationary chuck. Unlock ram on hexagon turret and turn spoke wheel to move
    stationary chuck away from work piece.

CONFIDENTIAL
SECURITY INFORMATION

RESTRICTED DATA
This document contains restricted data as defined
in the Atomic Energy Act of 1946. Its transmittal
or the disclosure of its contents in any manner to
an unauthorized person is prohibited.

Top secret instructions for using a turret lathe for milling uranium slugs at
Alba Craft, Inc., 1953. (Department of Energy document. Provided by the
Oxford Citizens for Peace and Justice archive.)

Much work would have to be accomplished, and it would have to be
accomplished with great expediency. Such expediency would prove

worthwhile for the subcontractor. Financial incentives for Albaugh
would increase over the five-year span of contracts. In 1951, the subcontractor was awarded $60,000, an amount that grew steadily to total
more than $500,000 per year for services rendered as the volume of
business increased.

The contract emphasized the sensitive nature of the work to be
performed.

> It is understood that disclosure of information relating to the
> work contracted for hereunder to any person entitled to receive
> it, or failure to safeguard all top secret, secret, confidential and
> restricted matter that may come to the Subcontractors or any
> person under its control in connection with the work under
> this subcontract, may subject the Subcontractors, its agents,
> employees and sub-contractors, to criminal liability under the
> laws of the United States (see the Atomic Energy Act of
> 1946).[5]

The penalties as stated in the AEC Act of 1946 ranged from
$10,000 fines and imprisonment up to life, even to death.

Interestingly, the contract never uses the term *uranium*, referring
instead to the radioactive element as "material." However, under
Section 9, Disclosure of Information, a significant clarification
occurs: "The term 'restricted data' as used in this paragraph means
all data concerning the manufacture or utilization of atomic weapons, the production of fissionable material, or the use of fissionable
material in the production of power." Furthermore, the document
includes two sentences insisting the Subcontractor "take all reasonable steps and precautions to protect health and minimize danger
from all hazards to life and property." At the same time, the contract declares the

> Government and the Contractor, their officers, agents and
> employees, harmless from any liability of any nature or kind, for
> any account of any claim for damages to Subcontractor's prop-
> erty or injury or death to personnel of the Subcontractor which
> may be filed or asserted as a result of the services performed
> under this Subcontract.[6]

While precautions for health and safety are mentioned, the drivers
of the enterprise are not held liable for loss or damage, injury, or
death.

Subcontractor Albaugh prepared to start milling uranium slugs by first building the work facility on West Rose Avenue. The original concrete block structure contained about 6,000 square feet of work area. It was expanded a couple of years later to 8,000 square feet, with interior walls to partition the different tasks. The building had no automatic ventilation system but did have several outer doors and some openable windows. Fans were often directed in and out of the building to offer some relief from the southern Ohio heat and humidity. Fans helped remove the milling dust, too, exhausting it directly into the yards, grounds, and streets surrounding the building.

A DAY IN THE LIFE INSIDE ALBA CRAFT, INC.[7]

In 1952, there were about 20 or 30 pieces of machining equipment in the building (lathes, drill presses, milling machines, grinders, filers, and saws). This increased to about 50 machines by 1955. It was obviously a busy place, and the operation was ongoing 24 hours a day for much of its existence. One trip report written by NLO in 1956 includes an estimate of "personnel working 84 hours/week."[8] Lloyd Singleton, who worked at the facility for several years, said, "We worked in 12 hour shifts seven days a week from 8am to 8pm and then others worked again until 8 the next morning."[9]

A typical day at Alba Craft began with the evening shift. The "material" arrived every evening in trucks or by rail cars that were sometimes labeled as containing radioactive materials, sometimes not. They contained heavy wooden boxes about "three feet long and 16–18 inches high on each side."[10] The boxes contained leadlike cylindrical slugs, each about six or seven inches long and ranging from less than a half to one inch in diameter. The material was malleable, but it was very abrasive and quickly wore down the bits used to drill holes in it, so Albaugh invented a special carbide-tipped blade to work the material. Some believe this innovation helped him to gain the initial contract with NLO.[11]

The workers used the various machines to hone and shape the slugs into rods suitable for the large uranium-processing factories. The work entailed grinding, milling, and cutting the slugs until they had "a nice slick finish" according to Lloyd Singleton.[12] Once the slugs became polished rods, holes were drilled through them. During the milling and drilling operations, the slugs would get red hot and even burst into flame. To douse the flames, workers poured or dipped the

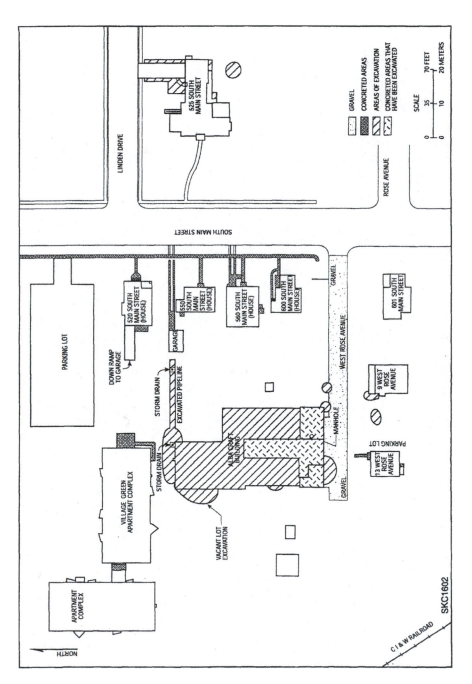

Drawing of Alba Craft, Inc., site showing areas of proposed excavation needed for cleanup, 1993. (Department of Energy document. Provided by the Oxford Citizens for Peace and Justice archive.)

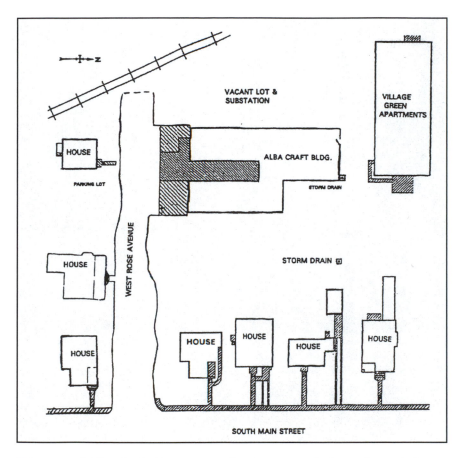

Diagram of Alba Craft, Inc., site indicating storm drains and proximity to railroad tracks. (Department of Energy document. Provided by the Oxford Citizens for Peace and Justice archive.)

flaming material into a milky liquid continuously. Streams of the liquid "perpetually flowed from every machine and onto the uranium as it was being drilled," one account relates. "The coolant made the air [in the facility] misty."[13] Sometimes the flaming material would fly off the machines, escape into the air, and land on the concrete floor. The workers would scoop up the flaming metal and place it in cans filled with coolant. The coolant was apt to boil, so to calm the boiling, workers often "kicked the cans to change the configuration of the uranium and stop it from burning."[14]

Such make-do techniques were common at Alba Craft. Working with uranium was new stuff, and even Albaugh was just figuring it out.

Photograph of Alba Craft, Inc., building prior to demolition, March 6, 1993.
(Photograph provided by Linda Musmeci Kimball and the Oxford Citizens
for Peace and Justice archive.)

Clyde Valentine, a military veteran and student of Albaugh's hired in
1952 said, "Lots of trial and error went on in the shop. We experienced
the flaming of the material in dramatic ways. I think even Bill was surprised at the way the material reacted at times."[15]

When the coolant in either the small cans or larger 50-gallon drums
became thick and unusable, the containers would be sealed. Holes were
drilled into the lids of the containers, however, to keep the cans from
exploding. One worker recalled a time when one of the big drums
caught on fire and they couldn't put it out. They called the local fire
department, which arrived and managed to extinguish the blazing drum.
"Albaugh wasn't there when that happened. When he showed up, he was
quite upset that the fire department had been called. He didn't want
them to know what we were doing in the shop, I guess,"[16] said one former worker. The containers were usually given back to the trucks that
came nightly. But sometimes, when they ran out of containers, the coolant was merely poured into the storm sewer grate adjacent to the shop.

Workers were supposed to wear long-sleeved white coveralls for
every shift. Not only were they to wear the coveralls and high-top

boots, but each day meant a change of coveralls and boots. On the front of the coveralls, the workers were also supposed to wear a badge of shiny metal with a plastic front. "We didn't know what the badges were for," one worker said, "but later we realized they were probably for measuring radiation. We never heard about any radiation levels, though."[17]

In this intense, piecework environment, there wasn't much time for anything but milling, grinding, and drilling the material. "When you walked away from your machine, the next guy would walk up and take over,"[18] reports a former employee. Workers often ate lunches and dinners in the building during cold weather and outside in the summer. One worker's wife recalled the protocol for bringing lunch to her husband. She was not allowed to go inside, so she handed it to him through the door. Workers maintained a tight work schedule seven days a week to meet the quotas and earn a decent wage. Some workers felt the money was good, others thought it mediocre. One worker estimated that he made about $1 per hour.

The workers, who numbered about 40 by 1955, primarily came from the greater Oxford area, but many were recruited from parts relatively far away. One area that provided a number of workers was Totz, Kentucky. Why they came from the town of Totz, almost exclusively, is not clear, although word of mouth seemed to have been a common mode for recruiting workers to Alba Craft. Some have speculated that Albaugh favored men from that area in the hopes that it might inhibit local gossip.

A DAY IN THE LIFE NEARBY ALBA CRAFT, INC.

Working at Alba Craft during its five years of existence tells one story of the uranium cottage industry. One hundred or more people were involved as full- or part-time workers in that period, and they reflect several attributes of the atomic mindset: They were loyal patriots, dedicated to the clandestine nature of the atomic enterprise; they were willing to work hard to meet deadlines, and they even reflected with skepticism occasionally. In the end, many workers believed in their work; however, they also remained uninformed about the potential dangers and risks associated with their jobs. Alba Craft workers were like many others carrying out their lives in the atomic development complex during its formative period.

A critical difference between Alba Craft and the rest of the greater atomic development complex was the physical place of the site itself.

It was in a residential neighborhood filled with everyday citizens, their children and pets, gardens, and backyards where they relaxed, ate, read, and did all the other things people still do in our privately owned habitations to pursue happiness. These backyards surrounding Alba Craft seamlessly and unknowingly integrated with the uranium-processing environment.

I have had the good fortune to talk with some of these residents. Those who are no longer alive often left behind print documents, recordings of oral interviews, and photographs of the site. Together, the stories are a fabric of witness accounts and archival records of life in the backyard of Alba Craft.

A variety of messengers describe life in the Alba Craft neighborhood. Although residents are the likeliest storytellers, one "agent" best frames their stories: water. Water, necessary and ubiquitous, tells history from particular metaphorical and literal perspectives.

Water was integral to uranium processing at Alba Craft. It cooled the machines and doused the flaming metal when it became too hot to contain its own energy. The milky water used to cool the product ran onto the floor, splashed onto the workers' boots and coveralls, and eventually flowed out the door and onto the ground. Clear water, sprayed on the lathes and drills frequently, kept the machinery clean and in good working order. This water also flowed into Rose Avenue and the surrounding neighborhood. Nature contributed as well. Rain poured onto the roof and property of Alba Craft, washing off the products and byproducts of the uranium industry. This water was common to all families in the community.

The Robinson family lived at 550 South Main and then moved to 552 South Main between 1952 and 1971. They were blessed with six daughters who spent their formative years on Main Street—Terry, Gail, Carol, Kelley, Peggy, and Amy. The water that ran through their neighborhood was an endless fascination for all of them. Carol held the distinction of being the mud pie queen, according to her sister Terry. She and her sisters and friends played in the "pretty water" whenever they had the chance. They made a secret hideout in the bushes growing along the perimeter of a spill basin adjacent to the Alba Craft property and pretended to fish in the basin or splash in the water when the weather was hot. "There was stagnant water everywhere," Carol wrote. "We played, and we had a ball."[19]

In 1959, when I was six years old, I had my first thyroid surgery. The surgeon explained to my parents that I had a Herthel-celled tumor with a cyst on it that was removed from the right lobe of the thyroid. In 1968 at the age of 15, I had my second thyroid surgery. When I was 18, I noticed an enlargement on the right side of my thyroid. The surgeon told me "not to worry about it" and said that no surgery was needed at that time. At 26, I saw an endocrinologist. He ran a battery of tests and said he did not want to do more surgery as scar tissue had formed and the nerves in my neck could be damaged or destroyed and my voice could be damaged. I have a goiter now that makes it difficult to swallow.[20]

Five of the six Robinson sisters have similar thyroid problems: colloid cysts, thyroidectomies, goiters, malignant growths of the thyroid, and other medical afflictions. When ingested by humans or animals, radioactive material, like water, moves until it finds a resting place—often in the thyroid gland. Water will evaporate, but radioactive materials stay. The alpha particles of radioactive substances stay in one place and continue their ongoing bombardment for years. The half-life of U-238 is 4.46 billion years.

In the Alba Craft neighborhood, water sustained vegetation in the gardens and berry patches in the backyards and on the uranium-processing site itself. "All the families had large vegetable gardens—tomatoes, broccoli, asparagus, peas … fruit trees were abundant, too,"[21] Rose Avenue neighbor Yerevan Peterson reminisced. "Mrs. Robinson and I had planted raspberry bushes at the back of our properties and the girls remember while playing they would eat berries right off the bushes."[22] Raspberries, being the aggressive plants they are, even grew along the walls of the Alba Craft building, where neighbors and workers alike picked and ate juicy handfuls for a snack or part of a meal.[23]

Water also had an indirect effect upon the produce of the neighborhood. It can be humid in the long, hot summers of southwest Ohio, and before air conditioning was common, electric fans would cool the workers in the moisture-laden air. Through the open shop windows, the milky mist and humid air of the facility would blow into the residents' backyards. No one knows how much uranium dust was in that ventilated air, but there most certainly was radioactive dust coating nearby plants.

The basements of the neighborhood were commonly visited by water, too, especially when rains were heavy. "One night I heard water

running and I looked out the door, the back door," said Alba Craft
neighbor Darlene Burch of 550 South Main St. "I couldn't find it
and finally I looked out at the factory and a river was just running.
We dug ditches to divert the water away from the house and that
helped some, but the basement would still gather water."[24] At the
time, they thought that it was just the usual kind of water that commonly flows through Oxford basements during torrential downpours.

Yerevan Peterson remembers often having to call on the expertise of
a plumber who used a sump pump to empty the basement of water.
Not uncommonly, though, Yerevan and her husband relied on their
own bucket brigade.

> Spiro would go down the basement stairs and wade in the water
> and fill the bucket. He wore boots, but no gloves as there seemed
> no need to do that. He would hand me the buckets and I would
> carry them outside and dispose of the water into the sewer grate.
> Spiro began to develop a painful rash on his hands and lower
> arms during this time. The rash remained with him for the rest
> of his life. He died in 1992, a year before the Alba Craft uranium
> processing business was revealed. I immediately suspected what
> he never knew—the water was likely the cause of the rash.[25]

A colleague of Spiro's in Miami's Department of English confirms
that Spiro

> did indeed have issues with the skin on his hands. Spiro was a
> Defoe scholar and often had to handle rare documents. He always
> wore cotton gloves to protect the pages. Although using gloves
> for this purpose is common, Spiro was especially diligent in doing
> so. He was afraid that the rash, which would sometimes weep,
> might stain the documents even if they weren't particularly
> rare.[26]

Many of these well-protected documents now reside in a special
collection that bears his name in the Miami University Library.

A most poignant story of the Alba Craft water is one that Yerevan
shared with me as we had lunch one day.

> The rash on his hands was really quite painful. He put creams and
> lotions on them, but it didn't take care of the pain. It hurt to have
> hands squeezed, so he often apologetically avoided handshakes.

He also avoided holding hands. We had always held hands, but after the rash developed we didn't anymore. I was always somewhat envious of couples that would walk hand-in-hand.

Walking hand in hand, such a simple human act that can at once vanquish anxiety and bring forth happiness, is itself vanquished from the lives of some who have experienced the power of the atom: a familiar story told time and time again.

NOTES

1. J. Bercaw et al., *IES Public Service Report*, 1–2.
2. G. Udell, Atomic Energy Act 1946, 38.
3. National Lead of Ohio (NLO) 1952–57 declassified contract documents. These declassified top secret documents reside in the archives of the Oxford (Ohio) Citizens for Peace and Justice archive.
4. Ibid.
5. Ibid.
6. Ibid.
7. Neighbors and Workers, 1994 WMUB broadcast, *In Alba-Craft's Shadow*, describing life in and around the site. This section includes a number of stories from the WMUB broadcast.
8. Ibid.
9. Ibid.
10. J. Bercaw et al., *IES Public Service Report*, 31. Worker interviews informing the IES report are included throughout this section.
11. Ibid.
12. Ibid.
13. Ibid.
14. Ibid.
15. Ibid.
16. Ibid.
17. Ibid.
18. Ibid.
19. Transcript of Public Hearing Oxford, Ohio, April 21, 1993.
20. Ibid.
21. Y. Peterson, 2009, 10; conversations with the author.
22. Ibid.
23. Ibid.
24. Public Hearing, April 21, 1993.
25. Y. Peterson, 2009, 10; conversations with the author.
26. F. Jordan, 2009; conversation with the author.

PART THREE

Confronting the Atomic Mindset: Community Action, the Rhetoric of Nuclear Power, and Fukushima

The final section of *Romancing the Atom* addresses the question, Once the atomic mindset has taken hold and events unfold in which atomic development compromises our environment, health, and way of life, what can be done to confront the mindset and work for change? Further, can individuals and small groups of people counter the "set" of the mindset and foster productive actions for the good of communities and, ultimately, humankind and all of the other residents—faunal and floral—of Earth? There are no clear answers to these questions, but the first two chapters provide examples of how confronting the atomic mindset can make a difference and how such confrontations can be accomplished. The section concludes with a chapter on the events surrounding the 2011 tsunami and nuclear disaster in Fukushima, Japan.

Chapter 8 continues the Oxford, Ohio story when the secret uranium milling operation is discovered in the 1990s and the citizens of the town confront the government over the task of cleaning up the radioactive debris that had festered in the soil, the water, and the machine shop for nearly forty years. The Environmental Protection Agency and the Department of Energy had both designated the site for cleanup, but the date for it to begin was more than a decade in the future. Many people in the town did not want to wait—the prospect of continuing to live literally with the site in their backyards was just not good. Over a three-year period, through a concerted effort by several committed citizens, the cleanup was accomplished much more quickly and thoroughly than if there had been no grassroots push to make it happen. The chapter includes personal narratives of those involved, documents from the government archives, several photographs of the cleanup itself, and most importantly, excerpts from the town meetings that told the stories of those affected by the milling site and their successful calls for action.

Chapter 9 continues our journey through the atomic mindset by coming to the present day, where we are now called upon to confront the questions surrounding nuclear power development. For more than three decades, the construction of nuclear power plants in the United States has come to a strict halt. Fears wrought from accidents and events such as the Chernobyl nuclear power plant meltdown and the nuclear accident and near meltdown in Three Mile Island, New York, raised fears in the United States (and some other nations) that slowed the construction of power plants. Currently, however, growing concerns among politicians, media pundits, corporate entities, and public organizations regarding the high monetary costs of petro-chemical energy production, coupled with the specter of climate change and greenhouse gases, have reopened the arguments for nuclear energy development.

The budding renaissance of the nuclear power industry has been labeled by some as "nuclear green" and is built upon, this chapter argues, a faulty set of arguments. Calling them "the green enthymeme," the chapter concentrates on the language of nuclear green proponents and the power that their language has fostered for revived construction of nuclear power plants. Drawing upon statements by politicians, media spokespeople, energy experts, and other outspoken advocates for nuclear power, the chapter explores how the advancing of nuclear power by certain powerful individuals and institutions has turned the tables from fear and skepticism toward an unreflective optimism that, in turn, reflects our romance with the atom once again.

Chapter 10 was literally a surprise in that the horrific events of Fukushima, Japan in March 2011 occurred after the original manuscript of this book had been drafted. This chapter adds, in a darkly serendipitous manner, to the stories of our romance with the atom. The unveiling of the events that occurred during the first days following the enormous tsunami and Dai-ichi nuclear power plant meltdown are drawn from accounts that are only now, more than a year later, becoming available due to the high levels of radioactivity, the extensive damage to the power plant and the city of Fukushima, and the secrecy (yet again) that limits the flow of public information about the disaster.

CHAPTER 8

What's a Community to Do?: Alba Craft, Inc., Part 2

Radioactive waste at Hanford alone would "cover an area the size of Manhattan with a lake forty feet deep."[1] Shockingly large amounts of radioactive wastes remain at Hanford and at Oak Ridge, Rocky Flats, Savannah River, Los Alamos, Fernald, and a host of other North American facilities. This we know. What we cannot fully account for is the array of secret sites of atomic pollution. By 1993, the Department of Defense (DOD) had in fact identified more than 11,000 polluted sites owned by the armed services, with more coming to light by way of secret documents declassified throughout the 1990s and continuing into the present day.

Beyond the contaminated lands owned by the armed services are those contaminated sites located in communities—on private and public lands. Alba Craft is one of these. It is merely one of possibly hundreds of sites that have been identified or may be identified someday in the future as the history of atomic development is pieced together. And wherever sites are discovered, cleanup should follow.

Cleanup is a key word of this chapter, and, as we will see, cleanup involves more than just earth-moving equipment, workers in protective clothing, and the problems of what to do with all of the radioactive debris. It also involves communities: the humans, wildlife, vegetation, land, and other things that constitute life as we know it on planet Earth. Further, communities intersect in numerous ways that create a complexity of interactions: What happens in one community can change and, in the most extreme cases, forever alter some communities. Radioactive cleanup is, in the most blatant terms, a community event that affects everyone and everything in the community.

SUPERFUND AND FUSRAP: SOME BACKGROUND ON TOXIC CLEANUP IN THE UNITED STATES

As industrial manufacturing devolved during the second half of the twentieth century, Americans gaped at the unsightly remains of early industry: pollutants from the mining, agriculture, steel, petrochemical, and other industries needed to be cleaned up. In 1980, as a response to major pollution sites in the United States, the government created the Comprehensive Environmental Response, Compensation, and Liability Act (CERCLA)—commonly known as Superfund—to remediate lands requiring long-term and intensive cleanup efforts. Superfund cleanups can focus on radiation contamination but also on other types of massive pollution, such as chemical and mining wastes and even just plain city or county landfill waste. Superfund projects often cost billions of dollars and take years to complete, if they even can be completed. For instance, the Rocky Flats nuclear weapons production facility outside Denver, Colorado, contained such large amounts of high- and low-level radioactive products that, despite major Superfund decontamination work, it is considered to be a National Sacrifice Area by the Department of Energy (DOE): an official designation for lands so heavily contaminated by radiation, chemicals, or other hazardous pollutants that they likely will never be totally remediated. Such lands are considered by the federal government to have been sacrificed for the national good. In some cases—such as around Hanford, Washington, or Los Alamos, New Mexico—these lands will never be used for any purpose other than as hazardous waste sites. However, some attempts to reuse these lands are being attempted, such as 4,000 of Rocky Flats's original 6,500 acres being named a national wildlife refuge in 2007 and placed under the auspices of the Department of the Interior. Time and the watchful eye of many wildlife scientists will tell if the wildlife community can once again flourish.

In addition to Superfund, the U.S. Congress created the Formerly Utilized Sites Remedial Action Program (FUSRAP) in 1974 to help identify and clean up sites specifically related to early atomic weapons development. By 1994, the DOE had discovered more than 400 potential FUSRAP sites and determined that at least 44 of these sites in 14 states called for cleanup. The criteria used to determine the need for a FUSRAP cleanup are essentially based upon radiation readings taken at the suspected sites. One notable criterion states that such sites

Table 8.1 Radiation Exposure Levels from Everyday Activities

Source	Exposure
External background radiation	60 mrem/yr, U.S. average
Natural K-40 radioactivity in body	40 mrem/yr
Air travel round trip (NY–LA)	5 mrem
Chest X-ray effective dose	10 mrem per film
Radon in the home	200 mrem/yr (variable)
Manmade (medical X-rays, etc.)	60 mrem/yr (average)

Source: Health Physics Society website, http://hps.org/publicinformation/ate/faqs/radsources.html. Used with permission.

may not expose humans or animals to doses of more than 100 mrem per year beyond "background radiation." Background radiation is the amount of radiation that an average U.S. citizen is exposed to each year, estimated to be about 300 rem/yr on average. Some of this exposure is due to natural environmental causes, such as sunlight, but it also includes radiation we encounter using everyday technologies. Thus, as Table 8.1 demonstrates, people's yearly dosage can vary depending upon where they live, how much they travel by air, and how many medical treatments involving radiation they might undergo in a given year.

Once a FUSRAP site has been determined to contain radiation levels over 100 mrem/yr, a computer code called RESRAD[2] models the current and potential exposures for the site. Accounting for soil types, rainfall amounts, groundwater level, and other environmental characteristics of the site, RESRAD can hypothesize scenarios involving human uses of the site to develop worst-case situations. These are used as starting point benchmarks to determine the need for cleanup and the processes for carrying it out. Once RESRAD has done its job, a second modeling technique called ALARA (or as low as reasonably achievable) is implemented. The ALARA principle is used to develop a plan for reducing radiation exposure to well below the 100 mrem/yr goal.

FUSRAP is required by law under the National Environmental Policy Act (NEPA) to adhere to environmentally sound practices during cleanup and to allow for public participation in the cleanup process. As a FUSRAP public dissemination brochure states:

Throughout the entire remedial action process there are opportunities for public participation. A community relations plan is usually developed at the beginning of the process, and the public is asked to provide information about the site, identify options, and comment on DOE's evaluation of the options. State and local governments and property owners also are key participants in this process. State and local governments help suggest appropriate and acceptable disposal sites that DOE should consider for the wastes and ensure compliance with the applicable state regulations. Local governments help inform the public about remedial activities.[3]

The process for establishing the need for cleanup at a suspected FUSRAP site is apparently straightforward. The site is identified through historical records; it is then examined through various assessment instruments to determine the levels of radiation present. Following this, a determination is made concerning its priority for cleanup, a cleanup plan is developed that includes a number of stakeholders, and then the site is cleaned, followed finally by future monitoring requirements. Once all of this is accomplished, the DOE develops a Fact Sheet that is placed on its website for the public record (see Alba Craft Shop Site).

The DOE FUSRAP Fact Sheet implies a smooth and seamless process. This final document records a cleanup in two pages, references other, more extensive documentation, and reveals that the entire process spanned a relatively short time period, about three or four years. The story of how it actually came to be, however, is much more complicated. Let's see how a community worked to make it happen.

ALBA CRAFT SHOP SITE

LTS&M Requirements

Long-term surveillance and maintenance requirements for the Alba Craft Shop Site and its associated vicinity properties are as follows:

- Managing site records,
- Responding to stakeholder inquiries.

The following *are not required* at the site:

- Monitoring: on site or off site,
- Site surveillance or inspection,
- Site physical property maintenance.

Following remediation, DOE certified that the site and vicinity properties complied with applicable cleanup decontamination criteria and standards and released the properties for unrestricted use. There are no supplemental limits, institutional controls, permits, or agreements in effect at the site.

Background and Supporting Information

Location

10-14 West Rose Ave., Oxford, OH
(includes nearby vicinity properties).

Ownership

Private.

Operations

Uranium metal fabrication (machining) for U.S. Atomic Energy Commission prime contractor National Lead Company of Ohio, 1952 to 1957.

Contaminants

Uranium metal.

Cleanup Criteria

U.S. Department of Energy *Guidelines for Residual Radioactive Material at Formerly Utilized Sites Remedial Action Program and Remote Surplus Facilities Management Program Sites.* Rev. 1, July 1985.

Site-specific uranium-in-soil standard: 35 pCi/g (DOE memo, Wagoner to Price, July 1994)

Remedial Action

1957, 1994, 1995. Shop property: remediated building surfaces and equipment, soils.

Vicinity properties: remediated building surfaces, soil, sewer line (*Post Remedial Action Report for the Former Alba Craft Laboratory and Vicinity Properties*, DOE/OR/21949-387, August 1995).

Release Survey

August 1995 (*Post Remedial Action Report for the Former Alba Craft Laboratory and Vicinity Properties*, DOE/OR/21949-387, August 1995).

Independent Verification

April 1996, Oak Ridge National Laboratory (*Results of the Independent Radiological Verification Survey of the Former Alba Craft Laboratory Site*, Oxford, OH, ORNL/TM-12968, April 1996).

Use Restrictions

Unrestricted.

Institutional Controls and Enforcement

Not applicable.

Monitoring and Site Inspections

Not required.

Certification and Regulator Concurrence

DOE certification March 1996; Federal Register Notice of Certification (published November 26, 1996, in 61 FR 60097); Ohio EPA and Ohio Dept. of Health, March 31, 1995.

Agreements and Permits

None.

Records Locations

National Archives and Records Administration Federal Records Centers, Kansas City, MO, Suitland, MD, and the DOE–EM Records Room in Germantown, MD; Considered Sites fileroom at DOE–HQ.

Source: U.S. Department of Energy, FUSRAP Sites LTS&M Needs Assessment, March 2005, Doc. No. S0164900.

FUSRAP CLEANUP FROM THE GRASSROOTS UP

In early 1992, the DOE had listed the Alba Craft property as a potential site requiring cleanup and, by the summer of that year, had carried out a preliminary investigation of the site's radioactive levels. The examination demonstrated that there was radiation contamination on the property and that the motion toward cleanup was inevitable. The amount of radiation they discovered, though, was deemed not a health threat by DOE, even though the DOE manager of the cleanup, Dave Adler, would report later that the contamination ranged from 10 to 100 times normal background radiation levels.

City officials, the owner of an adjacent apartment complex (Village Green), and the current owner of the Alba Craft building—which had housed the machine shop and now was an embroidery shop, professional office, and food warehouse—were notified of the DOE analysis and an estimate that remediation would occur 15 or more years later. But knowledge of the radioactive site would not become public for about eight more months. Fear of the unknown is housed within the atomic mindset. Fear of what radiation actually might do to humans and the environment was helping to maintain the grip of secrecy on those who knew. But what if the public knew—then would panic ensue?

About seven months passed by quietly. Then in March of 1993, Daryl Kimball of the Physicians for Social Responsibility (PSR) in Washington, DC, discovered the FUSRAP map indicating that the Oxford, Ohio, site was on the FUSRAP cleanup list and that it required remediation. The discovery of the FUSRAP site by Kimball was serendipitous. He just happened to be working for the PSR, he just happened to be from Oxford, and he had a mother who was well informed about atomic development issues and was actively involved with atomic topics and the nuclear weapon facilities complex through her work with the Oxford Citizens for Peace and Justice. This back-door manner of being informed about the radiation problems on Rose Avenue did not sit well with many people in this Ohio town.

"I don't feel good about the way I had to learn about it," remarked Linda Musmeci Kimball only a couple days following the discovery. "We didn't learn about it from the city, the state or the federal government. We found out about it through PSR."[4] It took the reading of a public notice, written by the Oxford Citizens for Peace and Justice (OCPJ), at the next city council meeting to really bring the problem into the public fold. The city officials had not appreciated the significance of the DOE radiological survey the previous year, and they needed it interpreted in clear, direct language. In essence, members of the OCPJ and fellow interested citizens had blown the whistle.

The explanation in the public notice did bring quick response by the city. City Manager Jim Collard said, "The whole episode is offensive to the citizens."[5] Collard had only been on the job a short time, but he did know of the study the DOE had carried out the previous summer. He knew that the city had been told that there were no immediate health risks associated with the abandoned site. Nevertheless, he assured the people of Oxford that he would send a letter to the DOE urging that the site be placed high on the FUSRAP list for cleanup. "We will be conducting a lobbying effort to see that these things happen," he said.[6]

Others expressed dismay as well as frustration with the flow and content of information from DOE. Gilbert Pacey, a chemistry professor at Miami University, had bought the Alba Craft building and property in 1988. Subsequently, he and his wife opened a collegiate embroidery business in the building and rented several rooms to a food warehouse company and another space to a laboratory business. "If I had known about it, I wouldn't have bought it. Even though there's no health risks, I don't want to fight the legal battles. I haven't made any payments since we heard about the contamination." He went on to explain that the two tenants had moved out when they were informed by Pacey of the DOE tests in the fall of 1992. He believed that they moved out because of perceptions and the difficulty of explaining to employees that it was a FUSRAP site. "I'm not moving because of the health risks. It's just people's perceptions, and the bank is leery of places with environmental problems."[7]

Two university student residents of Village Green Apartments adjacent to the Alba Craft site were also upset, and not only with the DOE, but also with the owner of the apartment complex, Jeff Schroer, and the editors of the *Miami Student* newspaper. In a letter to the editor of the *Miami Student*, Leslie Kimball and Kimberly Mayne chided the student newspaper for not interviewing students who lived in Village Green and only interviewing Mr. Schroer. "Ground contamination levels were *100 times* (sic) DOE safety standards at the time of the radiological survey of summer 1992, not just 'barely above natural background levels' " (as Schroer had claimed). "If he believes the press blew this out of proportion, we would like to challenge him to live in our apartment for nine months; how would he like to wake up every morning with the knowledge that there is known uranium contamination not 50 feet from his apartment, and that this information has been kept from him (and maybe others)?"[8]

In just a few short weeks, the local press published several similar reports and editorials depicting the community's concerns and fears about the FUSRAP site. Community involvement was gaining momentum, but so was government participation. The site manager for the DOE's FUSRAP Division, David Adler, responded to citizen concerns, providing helpful yet, at times, confounding information. Adler had been the person responsible for telling the public that indeed, the contamination was up to 100 times the background level in some of the samples that were taken in 1992. He also had explained that the site had been a uranium-slug processing shop in the 1950s furnishing fissionable materials for Hanford and other AEC production plants. He also, however, downplayed the contamination to some extent by describing it as a dusting of "oxidized materials that are adhering to the walls and floors" and that some soil samples outside the building were contaminated. "We're talking about trace levels of uranium that can only be detected with sensitive instruments. The only way it could pose a health risk is if people ingest appreciable quantities," he said in a press interview.[9]

Adler discounted the idea that a full-scale characterization study would be feasible for the site. Daryl Kimball, on the other hand, had stated that the extent of the radiation contamination and potential health hazards would not be known "until a full-site characterization is conducted."[10] Adler countered that such a study was not in the DOE's plans and that the study done the previous summer "was a fairly detailed survey . . . it was not just a few [soil] samples. It was a few days worth of technicians surveying the site. The next step will be clean-up." Then Adler added, "I don't know when this will occur. There is not currently any funding set aside for this year or next for clean-up of that site."[11]

Following Adler's comments, local citizens mounted a conflict-of-interest charge, represented by the Oxford Citizens for Peace and Justice through their director, Linda Musmeci Kimball. "On matters of environmental safety and health issues, DOE officials are the ones who set the standards, do the studies and form the conclusions," she said. "They have a conflict of interest here."[12] Lisa Fernandez, a press secretary for the DOE, responded by saying that "We are an independent body. We do not own the property. We do not own the site. We are asked by the owners to come in and ... survey it." She went on to explain that the EPA sets the standards, and the DOE just follows them.[13]

The proverbial bureaucratic problem of where the buck stops was a cause for concern. Just as the Diné of the Southwest and the Bikinian people had experienced, the issue of who was really in charge and whether the federal, state, or local authorities would act in accord with the community's wishes was promising to stall the cleanup. Further, the DOE released a report on March 27, 1993, that four other FUSRAP sites had been identified in Ohio alone, many similar to Alba Craft: sites that were being used for other purposes now but had originally served as part of the extended atomic cottage industry of the 1950s and 1960s. Upon the release of this larger FUSRAP cleanup need in Ohio, David Adler of the DOE said, "Generally all the contamination is contained within the buildings in which the uranium was machined. It's all what we would classify as fairly low-level, but I'm sure some might disagree with that."[14]

And disagree some did. "The DOE is not in the health business. Other medical and scientific sources would not share DOE's confidence in the premature conclusion that sites of this nature pose no health threats,"[15] Linda Musmeci Kimball declared. On this, community activists would have to get to work. The DOE said it had no immediate funds for further study or cleanup and that it wasn't necessary anyway in their point of view. Members of the community took matters into their own hands. Within a month, David D. Fankhauser, professor of chemistry and biology at the University of Cincinnati, completed an independent analysis of the 1992 DOE radiological survey. Community activists organized and held a three-hour-long public meeting, one of several that would follow.

THE FANKHAUSER REPORT

Wednesday, April 21, 1993, was a red-letter day for the Oxford community. It began with the release of Fankhauser's analysis of the DOE's 1992 radiological survey report. David Fankhauser was a good choice, as he had already done radiological analyses of the nearby Fernald, Ohio, radioactive materials processing plant, one of the leviathans of the industry. He knew what to look for and how the DOE normally carried out its analyses, and he was well schooled in the history and research of human interactions with radiation.

Fankhauser analyzed the DOE's conclusions and subsequently offered alternative recommendations.

The DOE reported that some samples had shown 100 times the safety standard. But Fankhauser noted that "a much more detailed assessment of the level of contamination in the recesses of the drainage system, with particular attention to low sedimentation points, seepage cracks and other means of concentration and escape of the radioactive waste"[16] should be done. He added that

> contamination could have been carried blocks away or more. It is clear that substantial spillage occurred [at the site] because the contamination penetrates at least 60 cm. It is particularly ominous that these high contamination levels are outside the building, where neighborhood children at play might easily have contacted them.[17]

Inside the building, he also found contamination to be "of concern for the risk it poses to current personnel, and, perhaps more significantly for those who may have worked in the facility during the decades."[18] He noted that in terms of current usage of the facility, there were two "major clusters of elevated contamination: in the 'food stores' section now used as a warehouse, and the 'shirt press and shirt production' areas"[19] where clothing was embossed for sale to universities and schools, among other clients.

Drawing upon his knowledge of epidemiological studies conducted over the years, Fankhauser brought the impact of such levels of contamination to the fore. "It should be noted," he wrote, "that epidemiological studies conducted at Johns Hopkins Medical School have shown that exposure of a fetus to a *single* (sic) X-ray doubles the death rate for the infant. The incidence for mortality due to leukemia is tripled, and the incidence for mortality due to respiratory diseases is quadrupled."[20] He added that, "past exposures from the contamination are expected to have been greater than those currently measured because of the inadvertent cleansing and masking effects of housekeeping, remodeling and painting which have taken place over the decades since 1957."[21] To find out more about dispersal of contaminants, especially the underground plume produced at the site, he recommended that additional study was warranted, especially because it was in a residential neighborhood.

He concluded the report with eight specific recommendations. Three of them related to further surveys of the drainage systems, the larger site characterization, and one suggested looking for toxins other than radioactive components, such as chemicals used in the milling processes. He knew from his work with Fernald that rampant dumping of chemicals by workers was commonplace but often ignored by DOE and FUSRAP because they focused on radiation exposures. Another recommendation was to stop using the building for the storage of foodstuffs and that current workers should no longer use the facility. In addition, he said simply that "the building should be removed," carefully and completely, including the concrete pads of the foundation. The continued presence would only "be a stain on the healthy environment and reputation of an otherwise attractive college town."[22]

The final three recommendations, however, were of a different nature. The first called upon the City of Oxford to be an advocate for cleanup and for citizens to be involved in all of these advocacy measures. The second called for historical research in order to create a list of workers from the site, local neighbors affected, and their offspring. And in the third, he called for a chronology of the users and ownership of the site over the years. Most of these recommendations were carried out, and the final three were most certainly taken to heart by the citizens of the town.[23]

FUSRAP cleanup procedures dictate that the DOE work with the community to develop a plan not only for planning and conducting the cleanup but also for involving active community participation. This process of community involvement, if left to the DOE administrators and bureaucratic meanderings, could take a lot of time. Delayed community involvement for the Alba Craft facility, however, was not an option for Oxford citizens and local government officials. They wanted to get started quickly, so they did.

THE OXFORD CITY TOWN HALL MEETING

On the evening of Wednesday, April 21, 1993, the Oxford city government and OCPJ convened a public hearing at the middle school. The participants included the DOE's FUSRAP project coordinator, Dave Adler, independent analyst Professor Fankhauser,

OCPJ board members, the mayor and city manager of Oxford, and a few hundred citizens of the town in what would be a long evening of personal testimony, scientific explanations, and calls for action.[24]

After some perfunctory remarks by city and DOE officials concerning the present state of the Alba Craft building and the movement toward a potential FUSRAP cleanup, Dr. Fankhauser took the stage and overviewed his report for this wider public audience. He also spoke of his previous consulting with the Superfund cleanup of the nearby Fernald plant. He speculated that some of his interviews with machinists and other workers at Fernald could shed light on the Alba Craft site.

> I spoke personally with machinists from Fernald who said it was regular practice to take trichloroethylene, a very poisonous solvent and carcinogen used for degreasing machinery, and take it to the door and pitch it in the parking lot. I don't know whether that happened at Alba Craft. I wouldn't be surprised.[25]

He went on to talk about the possible history of children who had played with the soil and hung around the site.

> You can't tell me that children have not played in that area, if they didn't dig in the dirt, if they weren't exposed to that, I'd be surprised. You need to ask questions: What was the exposure rate of those children; who played in the dirt? Were children exposed? If a child inhales the uranium dust, if a child eats the soil, if he gets it on his hands and eats, he gets it inside his body. Those alpha particles will smash cells like no other radiation.[26]

He concluded with a call to action by the people of Oxford. "We need to extrapolate back to get some idea of what the exposure was to the workers, and neighbors. It's going to be your job to see to it that the building is properly taken care of. And that means getting rid of it."[27] His call would eventually become the focus of the citizens of the town, but for the next several hours at the town meeting, the focus became the testimonies of the many residents who had lived near and worked in the facility.

> My name is Rose Patton. I live at 520 South Main St. I have lived there for 27 years. My children have grown up in the backyard; they have made mud pies that they have eaten. I am speaking to

you this evening because I am concerned as a mother and for my neighbors. I speak to you about my children and my grandchildren who played in that yard and have all had respiratory problems. I also want to tell of my yard that was lined by beautiful poplars and silver maple trees, some 80 or 90 feet tall. In 1969, the leaves started to all fall in July. The leaves never came back and all of the trees were dead within a year. I went barefoot a lot because I enjoyed it. I contracted something that is still on my legs 22 years later. It has never been diagnosed. They don't know what it is. My daughter, Rebecca, is now 37 years old. She now has a goiter. I am here tonight because I am concerned, and I am making a formal request to have my home, land, and my water evaluated.

My name is Terry Robinson Fulton. I'm 40 years old. I was born in 1952 and I lived at 550 South Main until 1955. When I was three, my parents built and moved into the house next door to the south. I lived there until 1971. Two of my sisters are here tonight. Three other sisters have moved from Oxford. Two of them sent me letters to be read tonight. Alba-Craft was going sometimes 24 hours a day. It was a noisy place, machining or milling was going on, we didn't know what. Since our raspberries grew right next to the building, we kids usually ate them as we played in the backyard. Our environmentally-minded parents didn't use pesticides. We were never warned about eating these berries unwashed. There seemed no reason to do so. Thyroid problems are a constant nag for us six sisters. And although I have not had surgery, three of us have: Gail, Carol, and Kelley. Gail and Carol run a high risk of losing their voices if another surgery is performed. Another sister, Amy, has not had surgery either, but she undergoes needle biopsies to relieve pressure in her throat and investigate fluid itself surrounding her thyroid. Amy is 15 years younger than I am.

Terry then went on to read from a letter penned by one of her sisters, Carol Robinson McLaughlin. Carol was the self-anointed mud pie queen. She wrote of the pretty water that she used to make her mud delicacies.

When they built the Coach House Apartments, now named Village Green Apartments, there was stagnant water everywhere . . . we had a ball. I had my first thyroid surgery when I was six years old. The surgeon explained to my parents that it was a Herthel-celled tumor with a cyst on it they removed. In 1968, at the age of 15, I had my second thyroid surgery. They then removed half of the gland. In 1979, an endocrinologist ran a battery of tests and told me that additional surgery was a concern as nerve damage to my neck was possible and my voice could be affected. I now have a goiter which makes it difficult to swallow.

Carol's letter was followed by another read also by Terry, but this time from Kelley Robinson Hickey. "In the early 1980s I had a hemi-thyroidectomy to remove a tumor." Finally, the last of the sisters, Gail Robinson Marbut added to the long tale of surgeries and physical and emotional pain.

At age 12, I suddenly developed a large growth on the right side of my thyroid gland. The majority of the gland was removed. Sometimes the gland swells and makes it difficult to swallow and breathe when I lie down. The doctor tells me that it is too risky to operate anymore. Unless a biopsy reveals cancer, he won't operate.

The testimony continued with more stories similar to these: inoperable brain tumors, recurrences of cancers, additional thyroid maladies. The list goes on. But nearing the end of the meeting, one speaker who was not a resident of Oxford—Ken Crawford—offered a story that expanded the Alba Craft issues onto a larger stage. Ken was a resident of Fernald, Ohio, and had been involved with citizen advocacy more than 10 years earlier when that large radiation uranium processing facility had been deemed a Superfund site. His testimony spoke of the misinformation and outright deceit that had been perpetrated by the Department of Energy and National Lead of Ohio (NLO)—the contractor that had managed the Fernald leviathan. He spoke fervently about his experience and did not mince words.

This plant was formerly called the Feed Materials Processing Center. This was the first deception because of the red and white checkerboard sign on water towers and dairy cattle grazing on the site. We lived in a farmhouse across the road. In 1981, a Unus Geological Survey revealed that our well was contaminated with uranium and had 190 times the normal background for drinking water in our area. The DOE and NLO knew this but chose to sit on their duffs. The citizens of Oxford are now in a situation that parallels ours. If you left things up to the DOE and NLO, you still wouldn't know. This is history repeating itself all over again.

He explained that uranium contamination at Fernald was of unknown quantities and that the scientists and management at the facility operated through what he called "the SWAG theory"—a scientific wild-ass guess.

"No one knew" continued as a lead refrain of the atomic mindset. And deception was close behind.

The meeting lasted for nearly two more hours, during which time a host of other community members—students, professors, members of various organizations, government officials—continued voicing concerns over everything from health problems to property values to contamination levels, and so on. By the end of the evening, more than 30 people had spoken, and there was one clear common thread that emerged: Community members had to take charge and be advocates if anything was to happen quickly.

BUILDING A COMMUNITY RESPONSE

Responding to the radiological contamination at Alba Craft continued through the year. It involved deep commitment by a number of community members. An advocacy group named Residents and Alba Craft Workers (RAW) was formed to spearhead research on those who had ever worked or lived near the facility. RAW collected as much information as possible about former employees, what houses they occupied during that time, and where they currently lived. Their work was rewarded with the names and history of many who had been affected, eventually recording the background of more than 40 individuals. Some of them were deceased, but some were still available for interviews and even medical testing. While many of the respondents indicated they suffered from health problems, it was difficult to pinpoint a cause–effect relationship between the Alba Craft facility and these maladies. Nevertheless, RAW continued its work and prepared information to be used by government and cleanup officials, as well as by residents. People wanted to know, and RAW became a key source for these concerned citizens.

RAW, OCPJ (led primarily by Musmeci Kimball), and several local government officials formed a delegation that visited DOE offices and federal representatives in Washington, DC, to hasten the cleanup starting date. As of early June, the DOE had not committed to starting the job anytime within the next five years. Undaunted, the delegation organized a meeting in late June with Ohio Senator John Glenn,

U.S. Representative John Boehner, and Assistant Secretary of Energy Dan Reicher in which they argued for cleanup to begin immediately. Officials made a few tentative promises, but commitment to action on the cleanup was still apparently on the back burner.

News reporters were especially busy investigating similar stories. Between April 25 and May 5, the *Hamilton Journal-News* wrote of several other sites in the greater Butler County area, thus fueling fear and apprehension in the area about the dimensions of this once-secret enterprise. News stories continued throughout 1993 and into and beyond 1994.

Community activism in the Oxford area was beginning to yield results. In July, the DOE returned to conduct further testing for contamination of homes, properties, local schools, sewers, and a local landfill. Among other things, this survey showed that Eugene Albaugh's former residence at 525 Main St. was contaminated with uranium oxide dust. This disturbing finding prompted the current owners, Marilyn and Wayne Elzey, to push hard for a cleanup of the house itself. In this case, the DOE acted quickly. From October 27 to 29, the DOE removed wood flooring and other internal parts of the house, finding more contamination as they went. Once the Elzey house was "cleaned," there were calls from the community to have other nearby residences cleaned. The DOE claimed that this was unnecessary, as the other houses were not contaminated. Many citizens were skeptical, yet no other houses were ever cleaned, leading some to believe that the DOE testing was inadequate.

By November, the entourage who had traveled to Washington during the summer was getting impatient. Not only had no action been taken on the promises made, but mundane safety measures such as erecting fences and mounting warning signs around the Alba Craft site to alert passersby and children who might still play in the area had not been done. Further, health concerns of the local residents and former workers were held in abeyance. So, off to Washington, DC, they went a second time to press their case for DOE action.

This visit spurred action. On November 29, in a letter from the Senate Committee on Governmental Affairs, Sen. John Glenn petitioned the Agency for Toxic Substances and Disease Registry "to evaluate the potential human health effects of the radioactive contaminants stemming from past operations at the former Alba Craft location, including past and future exposures."[28]

The letter brought prompt action. In January 1994, a public health consultation was conducted in and around the Alba Craft site, resulting in a full report from the agency on July 8. The report contained a lengthy history of the site that provided few things not already known, but at least it was official documentation by the government of the problems that existed. Beyond the history, however, the report offered some recommendations that would eventually lead to an accelerated cleanup schedule.

Beginning in August 1994 and extending through January 1995, remedial actions were undertaken. The Post-Remediation Dose Assessment Report written by the Environmental Assessment Division of Argonne National Laboratories noted that the on-site buildings had been demolished due to the "presence of radioactive materials ... and all building debris and radioactively contaminated soil were removed." In addition, "a contaminated sewer drain ... was also excavated, along with the contaminated soil surrounding the drain pipe."[29]

All in all, about 2,100 cubic meters of radioactive waste were removed. The concrete rubble and cinder blocks from demolition of the building complex were processed into a soil-like consistency and, along with soil and other materials, were packaged and shipped to their resting place in the Envirocare facility in Clive, Utah. The 40-year-old uranium processing site was finished off—the physical aspects of it, that is.

Photograph of partially demolished Alba Craft, Inc., building, October 1994. (Photograph provided by Yerevan Peterson and the Oxford Citizens for Peace and Justice archive.)

THE EMOTIONAL AND PSYCHOLOGICAL AFTERMATH

The Alba Craft site was barren. A lot bordered by a chain-link fence on the east side, it remained a large, empty lot for a few years until a real estate developer won the permit from the city to build a small apartment complex. Otherwise, the remainder of the site looks the same today as it did in 1996. There are no signs posted, no warnings. It just looks like an extension of the field that stretches out to the west and north toward the old railroad tracks and elementary school past the ball fields.

What remains, however, is not just the empty lot but a much fuller landscape of human emotion, fear, and nagging questions. Even though no radioactive contamination is evident at the site, radioactivity is still present in the town. Incidentally, the former Albaugh house had been thoroughly cleaned inside, but traces of cesium-137 were discovered by the downspouts. This was an unexpected find. Concerned citizens speculated about how cesium-137—one of the most potently radioactive

Rock crusher used to pulverize all of the cement blocks that comprised the Alba Craft, Inc., building in 1994 and described in the interview with Musmeci Kimball and Yerevan Peterson. (Photograph provided by Yero Peterson and the Oxford Citizens for Peace and Justice archive.)

elements known—could possibly have been generated by the uranium processing of the 1950s. Cesium-137 only results from extreme heat. In fact, it is usually only created by the heat generated in atomic explosions. In fact, they wonder, was it even remotely possible that the Alba Craft facility could have produced such a volatile element, such excessive heat?

Here we once again feel how unsettling a romance with the mysterious atom could be. A letter dated August 1993 from the Department of Energy to the former owners of the Albaugh house, Wayne and Marilyn Elzey, confirmed that indeed, cesium-137 had been discovered below one of their downspouts. "The cesium is from fallout from atmospheric nuclear tests; it was deposited on the roof and concentrated near the downspout by the runoff of rainwater from the roof,"[30] the letter explained. So the cesium-137 was not from Alba Craft after all. It was an orphan of the nuclear testing conducted over the years many, many miles away and deposited randomly on earth. Obviously, then, anyone could have received the mysterious gift of cesium-137, one of many possible unclaimed and unwanted atomic orphans to be discovered on roofs and doorsteps or below downspouts of homes in communities around the world.

Men using plastic tarps to try to prevent radioactive materials and other hazardous waste from being washed away by rains, fall 1994. (Photograph provided by Yerevan Peterson and the Oxford Citizens for Peace and Justice archive.)

NOTES

1. D. Kimball, *Covering the Map*, 6.
2. RESRAD is short for RESidual RADioactivity. RESRAD is a computer software program. For a full explanation see: https://docs.google.com /a/mtu.edu/viewer?a=v&q=cache:-gYaS-Qj4aIJ:www.ieer.org/sdafiles/15-4. pdf+&hl=en&gl=us&pid=bl&srcid=ADGEEShfnFQQSIJh7w677VStWXH Td1CbEu5iVAeyzXh-0dECr5N_9GMgoI0qCXPyiJA_kHj7gu_aQXCgC8A5 Fh3XxVDYyxfCXTC7KvJXaHgqLtxOp9HDUJd813N5IKb7Q2MWfUt9 PhpM&sig=AHIEtbRYLtIu3wdRge5elhz-5tP8Iz4Gug&pli=1
3. Department of Energy Brochure on FUSRAP.
4. L. Wright, "Oxford Group," *Journal-News*, March 24, 1993.
5. I. Wright, "Cleanup Unites," *Cinci Enquirer*, March 25, 1993.
6. Ibid.
7. L. Wright, "Planning Move," *Journal-News*, March 25, 1993.
8. L. Kimball and K. Mayne, *Miami Student*, April 10, 1993.
9. J. Feiertag, "DOE Can't," *Journal-News*, March 26, 1993.
10. C. Nurnberg, "Contamination," *Miami Student*, March 26, 1993.
11. Ibid.
12. Ibid.
13. Ibid.
14. J. Feiertag, "DOE Marks," *Journal-News*, March 27, 1993.
15. Ibid., A7.
16. D. Fankhauser, "Independent Report," 1993.
17. Ibid.
18. Ibid.
19. Ibid.
20. Ibid.
21. Ibid.
22. Ibid.
23. Ibid.
24. Citizens, Oxford Ohio Town Hall Meeting, April 21, 1993. The testimonies included in this section are all derived from the official transcript of the Oxford, Ohio, Town Hall Meeting included in the bibliography.
25. Ibid.
26. Ibid.
27. Ibid.
28. Senator John Glenn, July 8, 1994.
29. Argonne National Laboratories, "Assessment Report," 1996.
30. W. A. Williams, Department of Energy letter to W. and M. Elzey, August 31, 1993.

CHAPTER 9
Nuclear Green and the End of Power

Every art and every inquiry, and similarly every action and pursuit, is thought to aim at some good; and for this reason the good has rightly been declared to be that at which all things aim.

—Aristotle, *The Nicomachean Ethics*, Book 1[1]

Back in the 1960s, the popular folk group the New Christy Minstrels wailed, "Green, green, I'm goin' away to where the grass is greener still." Also in this historic moment, a burgeoning nuclear power industry was singing its own green tune. Nuclear energy was on the make, promising to solve every known and imagined problem from sea to shining sea. It was singing its green tune from the boardrooms to the classrooms of America. Many can still recall the bewildering and vivid duck-and-cover newsreels schoolteachers showed impressionable students between episodes of long division and sentence diagramming. An often seen close-up from nuclear-age propaganda films was of a thumb and forefinger clasping a cylindrical pellet the size of a horse pill. Children were told in the voiceover that this one small pellet could power all kinds of complicated pieces of technology that they would need when they grew up. In the background, they saw muscular nuclear submarines, battleships, automobiles, and power plants arising in the present moment of the classroom newsreel. In the Sixties, the atom had become the magic bullet of our futures. A new day—one where the green grass would be greener still—was on the horizon.

Whether atomic energy has delivered a greener day is debatable. It exists now as a site for arguments from all sides of the atomic mindset. But how are these debates being waged, and are they worthy? How can ordinary people parse the many claims in the debates? For some insight into the nature of the current debates centering on "nuclear green," we'll need a short lesson in rhetoric.

Rhetoric is the study of language and its uses. As most readers realize, language can illuminate, explain, create wonder, and please, but it can also deceive, misinform, and harm. Rhetoric helps explain how and why language is used as it is on any given occasion. We sometimes hear the term *rhetoric* used pejoratively, as in "that's just rhetoric" to describe an utterance that is empty of meaning and designed to only serve the purposes of the speaker or writer. However, rhetoric has a nobler role: It is the careful study of how language actually functions and works in everyday life. It is in this second sense that I will use the term *rhetoric*.

Aristotle famously states in his *Rhetoric* that rhetoric is the ability to discover the available means of persuasion. Thus, rhetoric is a tool of persuasion and its attendant modes of argument. In making arguments, there are various ways to proceed, but two common structures pertaining to this discussion are those of the syllogism and the enthymeme. A syllogism is a structure of deductive argument and it consists of three main parts: a conclusion preceded by two premises. For example, "All men are mortal. Socrates is a man. Therefore, Socrates is mortal," demonstrates through a deductive three-step process that, in conclusion, Socrates is indeed mortal.

An enthymeme, however, truncates the syllogism to simply state—"All men are mortal. Therefore, Socrates is mortal." This implies that a common understanding (or shared premise) of Socrates being a man is already agreed upon by the hearers. A contemporary example of an enthymeme was provided by President Bush in May 2005 when he alluded to the September 11, 2001, attacks on the World Trade Center. Standing on the deck of the USS *Abraham Lincoln*, he declared victory in the Iraq War and said, "With those attacks, the terrorists and their supporters declared war on the United States. And war is what they got." The invisible piece of Bush's argument was the premise, "we all agree that Saddam Hussein was involved in the 911 attacks" (even though he apparently was not), which allowed and encouraged the president's listeners to immediately make, and believe, their inference—Hussein was behind the 911 attacks—without Bush having to state it explicitly. Bush cleverly used an implication to exploit his audience's willingness to be persuaded to believe in a villain, Saddam Hussein, and a hero, George Bush.

Using an enthymeme in arguments is allowable when there is a common understanding—a community of agreement—that makes

the middle premise unnecessary. This is a perfectly acceptable argu-
mentative move within the arts of persuasion, but it is not without its
problems, chiefly the one of injecting an illegitimate premise.

In the current debates over atomic energy, we have begun to operate through an enthymeme of "green" in that our atomic mindset is
willing to move from proposition to conclusion without the cumbersome middle premise. In fact, the enthymeme is a perfect vehicle for
a mindset; communal agreement is already present, so there is no need
to argue the particulars. Particulars are, in a mindset, just a waste of
time. The nuclear green enthymeme goes like this:

Nuclear power does not produce CO_2 emissions.
Therefore, nuclear power is green.

The communal perception that whatever does not (supposedly)
emit CO_2 is a universal good that allows politicians, scientists, industrialists, and the general public to begin accepting (and promoting)
the rationale that nuclear power is the most plausible solution to coal-
and petroleum-based electrical power plants. These arguments are
predicated on the nuclear green enthymeme.

Inherently expedient, an enthymeme can engender forgetfulness
and unreflective action. In the case at hand, actions that promote
nuclear power are motivated by needs and desires: for jobs, cheap
energy, and international markets. We are in fact currently experiencing the atomic mindset, through the nuclear green enthymeme, and
moving steadily forward unreflectively once again. Calls from across
the ideological spectrum to open doors to new nuclear power plant
development are growing nearly as fast as they did during the uranium
prospecting rush of 60 years ago.

Prospecting and its relatively slight disturbance to the environment,
however, is only the beginning. Once ore is found (and sometimes it
does involve more excavation than at other times), it must be mined.
To extract the raw metals from the earth entails more than simply digging, tunneling, and scraping. It also requires explosives. Sometimes it
means forcing caustic chemicals into the land and drawing the ore out
into processing units that "wash" the ore from the chemical soup
(a process called in situ mining). Following the mining, the ore must
be further processed, refined, hauled, treated again, and then made
into the various radioactive elements and products used by the atomic
and nuclear industries. Finally, the toxic remnants of the mining, the
trucking, the processing, and the radioactive debris must be discarded.

In the case of highly radioactive waste, we still have no official repository established *in the world* as of 2012.

Very real human health and safety risks are presented by the cycle of mining, in addition to all of the potential industrial problems associated with uranium prospecting, mining, and processing. We have grown quite comfortable in our culture assessing risk after the fact. This is also part of the atomic mindset. Even though we know a great deal about the effects of the entire atomic enterprise on the environment, animals, and humans, we just "let it be" and continue forward as if no history exists. As health physicist Dr. Karl Z. Morgan said in a 1983 interview after four decades of studying the health effects of radiation on prospectors, miners, and workers,

> Back in the year 1500 it was known that miners in the cobalt mines of Bohemia and Saxony were dying of the so-called miner's disease. And yet is hasn't been long in this country since we've had many miners working underground at levels as high or higher in radium and radon as existed in these mines 400 years ago. I think we'll be having many sad lessons before we learn what we should already know by now. There is no safe level of radiation exposure. So, the question is not: What is a safe level? The question is: How great is the risk?[2]

Unreflective forward movement toward nuclear power development might seem like the mindset of industries that want to garner profits from a new uranium rush or of individuals who want to cash in on another prospecting rush. And, in part, that is the case. For instance, in 2003, there were only 10 uranium claims made within five miles of the Grand Canyon. In 2009, there were at least 1,100 claims in that radius and another like amount in the five miles beyond that. Further, on April 27, 2009, the Bureau of Land Management (BLM) authorized new uranium permits for this same area—an area of public lands that had been protected by an order in June 2008 by the House National Resources Committee withdrawing the lands from potential exploration and mining. At this point, no mining has begun. However, the stage is set, and for a new play. We are accepting the hidden premise that nuclear power is not very burdened with all of this uranium prospecting, mining, and refining process. Instead, the enthymeme sounds like it is saying, "if nuclear is green, then it must be good. Nuclear energy is *the answer*." Nuclear energy is our hero or magic bullet. But we are forgetting that we've been fooled by heroes and magic bullets in the past.

So, who is arguing the nuclear green enthymeme? Many voices. Consider these statements from people spanning a variety of scientific, political, and mass media arenas. The italics in some of these statements are my own.

IN THEIR OWN WORDS: THE STORY OF THE NUCLEAR GREEN ENTHYMEME

With exceptions noted, quotes below are drawn from the Nuclear Energy Institute website in their "Viewpoints on Nuclear Energy."[3] I have italicized key passages to highlight examples of the green enthymeme and the rhetorical tools speakers employ to convince readers and listeners to adopt a mindset that won't question "the good" of nuclear power.

SCIENTISTS AND ENVIRONMENTALISTS SPEAK OUT

There's *no question* that [nuclear energy] is a clean way to generate huge amounts of electricity. There are *no emissions, no pollution*, and I think it is a very positive development.
  —Max Schultz, Senior Fellow, Manhattan Institute, April 2007

The fact that no nuclear power plants have been built in the United States in years is *a threat to all of us*. Nuclear power is needed to help meet the increasing demand for electricity, because it's the *only source* that can provide large amounts of power *without emitting carbon dioxide or other global warming gases*.
  —Dr. Nolan Hertel, Professor of Nuclear and Radiological
                     Engineering, Georgia Tech, July 27, 2007

We've *no option* but to use nuclear power.
                  —James Lovelock, atmospheric chemist and author
    of *The Revenge of Gaia: Earth's Climate Crisis and the Fate of
                    Humanity*, *The Observer*, May 2005

Sure, nuclear waste is a problem, but the great thing about it is *you know where it is and you can guard it*.

—Stewart Brand, Environmental Scientist and
founder, editor and publisher of *The Whole Earth Catalog*,
in the *NY Times*, February 27, 2007

If all the electricity you used in your lifetime was nuclear,
the amount of waste that would be added up *would fit in a Coke can.*
—Stewart Brand on TED.com, June 2009

Nukes are green.
—Stewart Brand, in *Whole Earth Discipline: An Ecopragmatist Manifesto*, 2009, p. 76

New nuclear generation, which has almost *no carbon* contribution
and *a tiny footprint* on habitat, must be significantly increased.
—The Boone and Crockett Club, "Climate Change Policy,"
August 2009

POLICYMAKERS SPEAK OUT

We have a new coalition . . . from the faith based community that
sees this [nuclear power plants] as a matter of Earth stewardship
and moral responsibility, to the nuclear industry, that knows they
are emissions-free . . .
—Sen. John Kerry (D-Mass), June 7, 2010

America's electricity is already being provided through the
nuclear industry efficiently, safely, and with *no discharge of greenhouse gases and emissions.*
—Vice President Dick Cheney, May 22, 2001[4]

The 103 nuclear power plants in America produce 20 percent of
the nation's electricity *without producing a single pound of air pollution of greenhouse gases.*
—President G. W. Bush, June 2005[5]

But the fact is, even though we have not broken ground on a
nuclear power plant in thirty years, nuclear energy remains our
largest source of fuel that produces *no carbon emissions.*

—President Barack Obama, announcing the first loan
guarantees at the IBEW Local 26 Headquarters
in Lanham, MD, February 16, 2010

... we need more production, more efficiency, more incentives,
and that means building a new generation of *safe, clean nuclear
power* plants in this country.

—President Barack Obama, State of the Union Address,
January 27, 2010

There's no reason why technologically we can't employ nuclear
energy in a safe and effective way. Japan does it, France does it
and *it doesn't have greenhouse gas emissions.*

—President Barack Obama at a town hall meeting in New
Orleans, October 15, 2009

Nuclear power is at the cornerstone of our *clean-energy* future.

—Gov. Jan Brewer (R-AZ), January 5, 2010

Cheap energy is essential to our economy, and we can get it *without
dirty water or dirty air.*

—Sen. Bob Bennett (R-Utah), August 14, 2009

We need *to jettison cumbersome regulations* that have stalled the
construction of nuclear power plants in favor of a streamlined
permit system ...

—Senators John Kerry (D-MA) and Lindsey Graham (R-SC),
October 10, 2009, in the *NY Times*

So why not build *100* new nuclear power plants in 20 years?

—Sen. Lamar Alexander (R-TN), July 10, 2009

We ought to go *totally nuclear.*

—Sen. Richard Shelby (R-AK), April 15, 2009

THE MEDIA SPEAKS OUT

The *greenest approach* of all is to gradually supplant fossil fuels
through an expansion of nuclear energy *as quick as possible.*

—*Worchester (Mass.) Telegram,* August 18, 2010

Now, with the oil disaster still out of control in the Gulf, it's even more important that nuclear power be given increased priority by the Obama administration.

—*Cape Cod (Mass.) Times*, May 11, 2010

Given that nuclear power produces essentially *no carbon emissions*, it's an appealing option for consistent and relatively clean generation.

—*The Washington Post*, February 20, 2010

That's one initiative of the President's we can firmly stand behind—a renewed commitment toward nuclear power.

—*The August (GA) Chronicle*, February 17, 2010

The U.S. needs more Nukes.

—*Richmond Times-Dispatch*, March 9, 2010

We're going to need nuclear.

—*Florida Times Union*, May 20, 2010

As these statements illustrate, the nuclear green enthymeme employs several major themes: nuclear energy creates jobs; it is environmentally viable; it produces virtually no pollutants; it is cheap; it is safe; it safeguards our way of life and our lifestyle. Driving these claims is the idea that nuclear power is not only needed, but that it is an imperative. "We must" rings like a town crier's bell throughout the nuclear power renaissance, announcing a new way out of our greatest environmental threat—climate change and global warming—through a technology that has been unfairly misunderstood. "We must" act now or, it appears, we are doomed. Like the general announced to the Bikinian people so many years ago, this is all for the good of mankind. So, let's get on with it. Period.

The arguments for nuclear power have become a story of sound bites—those brief phrases that generate excitement and catch our attention, even if they have no validity or basis in fact—or ethics, for that matter. The arguments for and against nuclear power are incredibly complicated, but the atomic mindset seems content to just accept the green enthymeme. "No CO_2" production sure sounds good, especially when we are so concerned about the encroaching pall of climate change and global warming. Climate change and global warming *are* real, and humans will have to come to terms with the generation of byproducts of our comfortable lifestyles that cause these problems.

But nuclear power is *not* the answer, as founder of Physicians for Social Responsibility Helen Caldicott bluntly states in the title of her book.[6] Nuclear power uses enormous amounts of petrochemical resources to mine, mill, transport, process, use, and then store the wastes that will be on and in the Earth for centuries to come. It is the whole nuclear cycle we should attend to and acknowledge, not just the sound-bite arguments that make untrue, half-baked claims. Instead of being intelligently skeptical of the power of the atom, we are being fed fear of what the world will be like without nuclear power.

If there is one thing that the stories of our romance with the atom can teach us, it is to be critically aware of what is being professed and then sent our way by the nuclear power debaters. In other words, we should ask some straightforward questions to help penetrate the atomic mindset. Here are a few of those questions to ponder:

- Is nuclear power green? It is neither greener nor more sustainable in terms of CO_2 production than coal- or petroleum-generated power. The amount of energy and power needed to produce the full cycle of nuclear power from beginning to end is enormous. Coal, petroleum, and natural gas are used in great quantities to mine, transport, process, and eventually create the nuclear fuel necessary to run the nuclear power plants. At present, there are no viable alternatives to reducing the CO_2 emitted during the entire fuel cycle. Nuclear power is, unfortunately, not a green magic bullet.

- Is nuclear power cheap? The entire development cycle is so costly from beginning to end that no private energy companies will begin the production of nuclear power without government subsidies and guarantees. The 2005 Energy Act, and more recent proposed subsidies of at least $15 billion by the U.S. government, have not garnered any takers on the deal. It is just too economically risky, so these private corporations want even more government backing to offset any risk they may perceive. Further, the nuclear power plant development companies are threatened by the smaller-scale efforts of alternative energy sources: You can put a solar panel on your home's roof or a windmill in your backyard. Unless you are a radioactive Boy Scout, you likely will not build your own nuclear power plant. A proliferation of

individuals and communities who generate their power independently does not sit well with large energy conglomerates. They
would rather prohibit this than be in the position of controlling
grass-roots power generation.

- Is nuclear power safe? If you have gotten this far in *Romancing the
  Atom*, this question should be a no-brainer. Nuclear power advocates—even those who once were staunch antinuclear environ-
  mentalists—such as Stewart Brand (Whole Earth Catalog
  founder) and Patrick Moore (Greenpeace founder)—make the
  claim that nuclear power fears incited by the accidents at Chernobyl and Three Mile Island are, in a word, overblown. Their
  sound-bite language on the subject is misleading, however.
  Those two nuclear power events have created a nuclear poisoning
  legacy for people within hundreds of square miles of those events
  that will last for generations.

- Is nuclear waste disposal a solvable problem? Most nuclear power
  advocates have a common point of agreement on low- and high-
  level waste: It is a large problem. However, in the rush to develop
  more nuclear power plants, advocates acknowledge the problem
  of disposal but believe that science will eventually find a way to
  solve the multiple issues of what to do with these long-term
  byproducts, even if it takes 100 years or so. And there are no places to store these wastes that any country can agree upon. Inter
  the wastes in concrete and steel containers, bury it under miles
  of earth, jettison it into outer space, reuse it and make it less
  radioactive—these are all solutions with little basis in reality.
  The stuff is just too volatile, and we don't understand the consequences in the long term. What legacy do we want to leave future
  generations of plants, animals, and humans? This seems to be the
  larger question that is not adequately addressed in the sound bites
  and the "for the good of the planet and humankind" arguments
  currently being professed.

ACKNOWLEDGING FEAR: IN THE HEART
OF IT ALL

In the atomic mindset and the romance with the atom that have
evolved for more than 100 years, there have been two themes that
occur time and time again: secrecy and fear. Secrecy is still a strong

aspect of the mindset, although the last 30 years have seen some breaking down of the intense secrecy that governed atomic development after World War II and through the Cold War.

Fear, however, has not diminished but, rather, has taken on different guises and hues. Fear of atomic power, of course, has always been there in the sense of awe and wonder. Marie Curie and the makers of Undark certainly were enamored of the particular properties of the atom and its luminescent beauty. Their love affair with radium eventually led to their demise and foreshadowed the fear and panic that would come, but their fear was held in check by the promises radium might hold for the future in health remedies and industrial applications.

Public fear arose with the unleashing of the Bomb. This manifestation of the atom also engendered awe, but the awe quickly turned to fear of what had been wrought with the dropping of two atomic bombs on Japan. Fear of war and enemies brought forward the most immense military defense tool ever known, but the tool itself would become the thing to be feared. The fear of nuclear proliferation would ripple through the rest of the twentieth century and up to the present day, stoked by fear of communism by the United States and of democratic governments by the Russians. The hue of fear now is of rogue terrorists who might make "dirty bombs," or even greater explosive devices fueled by the almighty atom.

Fear of radiation in everyday uses has also arisen. X-rays, CT scans, radioactive iodine treatments for thyroid cancer, and a host of other medical and industrial uses of the atom have sparked worldwide concern among scientists, politicians, and the public.

Now, we have the fear of climate change and global warming that is bringing the atom clearly back onto the table. We fear for the future of the planet, and that fear evokes those numerous reactions to propel nuclear energy as the answer to that fear.

Has the atom come to resemble humankind, human agency so much that we believe the atom will save us? Or, as cartoonist Walt Kelly had his character Pogo declare, "We have seen the enemy, and they are us!" This seems more apt for understanding our love affair with the atom. It is in the heart of humankind that we must dwell— to paraphrase Albert Einstein—if we are to apprehend our romance with the atom.

NOTES

1. Aristotle, *Ethics*, 1.
2. Morgan, *interview* in Del Tredici, *At Work*, 134.
3. Nuclear Energy Institute, http://www.nei.org/resourcesandstats/ documentlibrary/safetyandsecurity/reports/what-policymakers-are-saying.
4. H. Caldicott, *Not the Answer*, vii.
5. Ibid., vii.
6. H. Caldicott. *Nuclear Power Is Not the Answer*, Melbourne: Melbourne University Press, 2006.

CHAPTER 10

Nightmares Revisited: Japan, Tsunamis, the Atom, and Ironies

At 2:46 p.m., on Friday, March 11, 2011, a 9.0 magnitude earthquake struck off the east coast of Japan. The quake was so powerful that even in Tokyo, nearly 250 miles from the epicenter, buildings shook and the land parted. "At first I thought it was a huge windstorm because the power lines were swinging so much," said John Boyd, a British journalist who has lived in Japan for 40 years.[1] But there was no wind, only the unsteady moving of the earth and buildings and people. The earthquake itself proved to have—in comparison to what would come—minimal immediate effect on the Japanese mainland. After all, Japan is accustomed to large quakes. They have been a common experience along the Pacific Ocean for millennia; the Japanese infrastructure and human response to earthquakes are well tested.

THE VIEW FROM INSIDE: LIVING THROUGH A MELTDOWN

About 150 miles north of Tokyo in the coastal city of Fukushima and much closer to the epicenter, workers at the Dai-ichi nuclear power station began to monitor the automatic emergency systems in place for an earthquake. The nuclear facility, comprising six reactors, had three (Reactors 4, 5, and 6) that were currently shut down for regular maintenance. Within five seconds, control rods inside the three active reactors moved automatically into place and halted all nuclear reactions in the cores. In addition, because the massive earthquake had brought down critical power lines all around Fukushima, the emergency electrical generators for the plant kicked in. Things were even going well in the spent-fuel pools on the top floors of all six reactors; water filled the pools and kept the highly volatile nuclear refuse from over heating. In fact, Reactor 1 spent-fuel cooling was progressing so rapidly that there was fear of overcooling that could damage the steel walls of the containment vessel. So the decision was made to stop the flow of the cool water. Everything, it seemed, was in good order.

Then, at about 3 p.m., a number of tsunami alerts were broadcast all along the eastern coast. The earthquake had created walls of water that were moving at rapid speed (more than 200 miles per hour) toward the city of Fukushima and the Dai-ichi plant. The first tsunami hit the power plant at a height of about 30 feet at 3:27 p.m., followed by several more that have been estimated as high as 70 feet. The barrage of water swept into the large turbine buildings and overwhelmed the instrument panels, knocking out power all through the six reactor sites. The sea water, combined with the power outages already caused by the earthquake less than an hour before, produced the ultimate nightmare and irony for a nuclear power plant: an entire blackout.

Inside the nuclear cores, the coolant water was boiling away, but without the power to "see" the interior through the instrument panels, the workers had to guess at the rate of evaporation. And, to make matters graver, the diesel generators that would usually have backed up the systems were rendered inoperable due to the flooding waters. Desperate to provide any power whatsoever, workers rushed to the plant's parking lots and took batteries from as many cars as possible and brought them into the hobbled plant. Using cables and other scavenged electrical components, they created a makeshift power system that provided enough power for the instrument panels to display the badly needed information about the bowels of the reactor containment quarters. The information at first appeared reassuring—the water level in Reactor 1 was above the minimum level necessary to keep the fuel rods from melting. By 11:50 p.m., the workers were able to construct a second car battery stopgap power unit for another set of instrument panels, but this time they showed that the water had evaporated, leaving the rods exposed to the air in the containment vessels. A meltdown was underway.

Radiation levels began to rise in the plant, forcing officials to prohibit entry into at least one of the damaged buildings, Reactor 1. Pressure was building in the containment vessels that held the now totally exposed radioactive rods. The temperatures would eventually rise to 1,300 degrees Centigrade, about 2,732 degrees Fahrenheit. At this heat level, the uranium in the rods was beginning to melt and create hydrogen gas that could eventually cause an explosion. Around midnight, the decision was made to release the pressure by venting some of the highly radioactive gas into the atmosphere, spreading the contamination into the surrounding land and neighborhoods. There was nothing else they could do.

But before they could begin the venting, the people who lived within 3 kilometers of the plant had to be evacuated; that perimeter was enlarged to 10 kilometers by sunrise, as radiation levels outside the nuclear complex were rising rapidly. Buses came and hauled people away from their homes, completing this task by 9 a.m., on March 12. The Fukushima residents, not unlike the Bikini Islanders 65 years earlier, thought they would be returning home in a few days.

The exhausting work of trying to restore power to the plant continued in the darkness, too, as did the tasks of trying to vent the accumulating hydrogen gas from the now-dry containment structure and getting water onto the melting rods. There was finally some relief as a power supply truck arrived outside Reactor 2 shortly after midnight. But this was as close as they could get to Reactor 1—the roadways were blocked with debris, and some of the roads had collapsed. Working by hand over the next five hours, the workers dragged with bare hands a 600-foot-long power cable weighing a ton across the debris-strewn grounds and into Unit 1.

At the same time, about sunrise, one fire truck finally made it to Reactor 1 and began to spray water onto the fuel rods. This proved to be immensely challenging and time consuming because the truck's storage tanks had to be repeatedly filled from large holding tanks nearby. It would not be until mid-afternoon that enough hosing was brought in to directly flow 80,000 liters of water into the reactor. This, however, was not going to be enough water, so another decision was made: Start pumping seawater into the structure. The operators had been reluctant to do this, as the salt water could eventually cause erosion of metals that comprised the containment area. However, such erosion would be a long-term worry, and the more immediate needs overrode the desire to preserve the future integrity of the structure. At this moment, it was more important to cool things down and, hopefully, avert an explosion from the quickly emerging hydrogen gas.

Fukushima Dai-ichi was by now a lost cause in terms of ever being operational again. This was of little concern to the workers, who had been trying to solve yet another major problem: the venting of the hydrogen gas to avert an explosion. This, too, had been challenging because the crews working on the power and water supplies had been evacuated. And there was an additional problem. The hydrogen gas release valves used were not operational from a remote place because these valves needed electricity in order to function. They would have to be opened by hand—potentially deadly work, as it would mean entering the now-evacuated Reactor 1 building and likely exposing workers to dangerous amounts of radiation.

Preparing to enter the Reactor 1 building, the workers took iodine tablets to lessen the radiation absorption into their bodies, donned protective suits with air tanks, and carried handheld dosimeters to measure their radiation exposure. Once the venting team received word that all residents had been evacuated from the 10-kilometer radius of the plant, they entered the Reactor 1 building about 9 a.m. As they walked through the flashlight-lit hallways, they saw that their dosimeters were exceeding the maximum safety level of 100 millisieverts, about the equivalent of receiving 25 mammograms or 15 chest X-rays at one time, or four times the amount of radiation a nuclear worker should be exposed to over an entire year. They made the decision to turn back short of their goal of reaching the valves. Another technique would need to be developed to vent the increasing hydrogen gas from the reactor, this time using a crane truck and a portable air compressor that could be used to blast the valve open.

The crane truck was used to lift an air compressor to the valve's location and—while metaphorically and possibly literally holding their breath—workers watched the pressure gauge in the car-battery-powered control room. By 3:30 p.m., the shots of air from the compressor had done the trick. Hydrogen gas was now spewing out of the Reactor 1 building into the air outside. The remote gauges began to show a drop in pressure, and there appeared to be a brief moment for celebration. And brief it was—six minutes after the valve had been opened, there was a flash of light, a spark, in the reactor building. The hydrogen that had risen to the ceiling of the building was more than the hobbled instruments had shown. In addition, unknown to the operators, other vent lines had been leaking all along, adding more volume to the explosive gas. A huge explosion in Reactor 1 blew the top off the building, and the walls fractured on all sides. The falling debris managed to sever the 600-foot power cable, the sole source of steady power to the reactor area, and even sliced the fire truck hoses. No water, no power, increasing radiation in the sir, more explosions; no place to hide. The workers fled the area of Reactor 1.

Over the next couple days, the flurry of activity to control a chain reaction of explosions in the other reactor buildings continued at a

24-hour-a-day pace, but to no avail. Workers, now struggling within a very radioactive hot zone, tried to get the power truck running and more water flowing into Reactors 2 and 3 that were now showing signs of pending explosions. On March 14, Reactor 3 was the next to blow despite heroic efforts by workers to keep the feeble energy supplies flowing. This event caused further problems within the now-crippled infrastructure and stifled efforts to maintain Reactor 2; Reactor 2 experienced internal explosions on the morning of March 15, followed later in the day with a great explosion in Reactor 4 that not only tore off the roof but also started a fire that promised to spread radioactive contamination even further. The only good news that day was the discovery that the spent fuel rods in Reactors 4, 5, and 6 had remained under water.

Over the ensuing weeks, there were more fires and some blasts, although not as great as those in the first several days. Workers continued to stop leaks of radioactive water that were making the environment surrounding the nuclear plant and the greater Fukushima community highly radioactive. The specter of radiation had cast its shroud over the power plant and would make the facility unusable forever. According to one expert analysis of the nuclear plant's future, in 10 years, the facility will have had all radioactive debris removed and the plant will be covered with a plastic covering to lessen further radiation contamination. In 100 years, there are two potential options. The first is to have had all of the buildings' materials, reactor cores, and other infrastructural residue dismantled and moved to a nuclear waste storage site. The second option is to entomb the entire facility in concrete, thus making the once-invisible radioactive shroud a very visible mausoleum for a nuclear power plant.

THE VIEW FROM OUTSIDE: LIVING
WITH THE AFTERMATH

Outside the Dai-ichi nuclear plant immediately following the rage of the tsunami, the world looked like it had come to a gruesome end. It was, in a word, apocalyptic. The town of Fukushima, as well as numerous towns up and down the coast, had been washed away. At least 20,000 people had been killed almost instantaneously. More would die of exposure and injuries sustained by the massive wall of water and the uncountable tons of debris that had rained down in the monstrous swelling of the sea. The people within 10 kilometers of the plant had much to deal with for now. The radiation that was escaping from the plant was a secondary worry; they first were tending to the wreckage and the unimaginable loss of life that was all around them. But that would not last for long; the radiation would soon arrive.

People who lived within 10 kilometers of the facility had already been evacuated in the early morning hours of March 12. The severity of the radiation contamination was evident in the air and water surrounding the plant, and there was fear that more would come. The first emergence of radiation had begun as soon as the tsunami caused some structural damage to the reactor buildings. But this was minimal, at least for now. Next was the intentional release of the radioactive hydrogen gas that, unfortunately, failed to prevent the explosions that released massive amounts of radiation into the air and water. The town of Fukushima was being covered by the invisible bone-seeking radiation, and the population would take on a new identity: radiation monitor.

As we have seen in nearly every chapter of this book, the monitoring of radiation levels has been part and parcel of the atomic romance for over 60 years. Whether it was workers in the uranium mining and processing industry, military personnel, medical personnel, or nuclear power plant employees, the constant vigil of keeping an eye on radiation levels has been an everyday practice around the globe. A chief question that surrounds all radiation monitoring is, What levels of radiation are safe? We are provided with numerous charts and graphs that portend to display what is an "acceptable dose" of radiation, whether it comes from natural causes such as sunlight or from artificial sources such as the nuclear development industry. These acceptable levels are developed by highly trained scientists, medical doctors, and statisticians who weigh the presence of certain ailments associated with radiation—most commonly, cancers—against the prevalence of that disease in a given population. The probabilities computed over some period of time, usually years, help these scientists to arrive at levels of exposure that are "acceptable."

Acceptable to whom, however, is another question. For instance, according to the International Atomic Energy Agency,

> The dose limits for [everyday] practices are intended to ensure that no individual is committed to unacceptable risk due to radiation exposure. For the public the limit is 1 mSv in a year, or in special circumstances up to 5 mSv in a single year provided that

the average does over five consecutive years does not exceed 1 mSv per year.[2]

For nuclear power plant workers, however, the levels of acceptable dosage are much higher. At Fukushima Dai-ichi, the upper limit of exposure was 50 millisieverts (mSv) to 100 mSv *per shift*. On March 18, just seven days after the accident began, the Tokyo Electric Power Company (TEPCO) raised the limit to 150 mSv. "This is a considerable amount of radiation," said G. Donald Frey, a medical physicist and professor of radiology at the Medical University of South Carolina. "The limit for radiation workers in the United States is 50 millisieverts per year, but we try to keep them to less than 5 millisieverts per year."[3] Given the context and the depth of the atomic mindset in that context, the rules of engagement are altered as needed.

On the other end of the radiation safety spectrum, we hear quite different notions of what constitutes acceptable risk. We already have heard Dr. Morgan, in a previous chapter, state, "there are no safe levels of radiation," based upon his several decades of study. This sentiment is seconded by Dr. Helen Caldicott, the founder of Physicians for Social Responsibility, when she remarked in the wake of the Fukushima disaster that "Doctors know that there is no such thing as a safe dose of radiation, and that radiation is cumulative."[4]

Again, what is a citizen or community to do? In Japan, as we have seen in other places, people took matters into their own hands to become informed in a manner they trusted: They became citizen radiation monitors. Kiyoko Okoshi took the lead in her home village, Shidamyo, about 20 miles from the Dai-ichi plant. Shidamyo is a rural town where farmers like Kiyoko have thrived for generations, working the land. Their lives had been what one might expect of a small rural village. "Our life was so lively when the four boys were running around the mountains in back of the house,"[5] she said. But her grandsons and daughter are no longer there. They fled because of radiation fears and their distrust of the government officials to tell them the truth about radiation risks. They had confronted the local officials, but there was no response, thus confirming their suspicions. So the family left, but Mrs. Okoshi stayed and began a grassroots effort to monitor the radiation. Her hope is that she will locate where the "hot spots" are, but, more importantly, that she will discover that the radiation is not so high. She wants her family to come home, and she wants them to be safe.

She purchased a dosimeter for about $625.00 and began her monitoring of the town. Her results were alarming. She found levels beyond what the officials had reported. Local officials viewed her with skepticism because she was a mere amateur in monitoring matters. But that reputation was soon overturned when a radiation expert, Shinzo Kimura, quit his position with the Japanese Health Ministry over concerns with the government's slow response to the crisis and came to the support of these citizen radiation monitors. He confirmed her readings, and this led to the town beginning its own monitoring program.

Others have taken up the work of monitoring their backyards and neighborhoods. In the city of Kawamata, the largest city near the Dai-ichi plant, officials have purchased and distributed 34,000 dosimeters to children between the ages of 4 and 15. Even in Tokyo, more than 150 miles from the plant, there has been growing concern that the radiation would find its way there in high levels. In Tokyo, at one seminar about the use of dosimeters, more than 250 people showed up, and officials had to turn some people away.

And in yet one more citizen-led effort, radiation was detected in Japanese-produced infant formula on December 6, 2011. The infant formula company, Meiji, announced a recall of 400,000 cans of the formula because of the presence of cesium-134 and cesium-137 in the product. The company speculates that the formula became contaminated from airborne particles, as they had "been diligent in checking radiation levels in the water, but had not taken enough care to filter for airborne radioactivity."[6]

THE IRONIES OF A NUCLEAR ACCIDENT

Several ironies mark the catastrophic earthquake and tsunami that rocked Japan's east coast on March 11, 2011. The first of these ironies is personal. Just as I had completed a draft of this book's manuscript and had written what was to be the final chapter of the book on the fallacies of the arguments surrounding the concept of "nuclear green," the tsunami struck and, on several fronts, a nightmare became a reality. "Nuclear green" had become "nuclear disaster." Another chapter was necessary.

The other ironies are on a much grander scale. First among these is the fact that this nuclear accident—one that may overtake some aspects of Chernobyl in the record books of nuclear power disasters—occurred in Japan. The country that was the target of the first atomic bombs was once again the target of the almighty atom. The country where several

hundred thousand Japanese citizens had died, or had been horribly
scarred both physically and psychologically by the deadliest weapon
ever unleashed, was now a victim of its own involvement with this most
powerful of energy sources. The Japanese government had chosen after
World War II to never produce an atomic weapon. Yet they trusted the
atom well enough to welcome it onto their soil in the pursuit of electric
energy. And this is not just in Japan; it is occurring across the globe.
Dozens of countries are in hot pursuit of developing nuclear power
plants (even though some countries have now placed them under a
moratorium due to Fukushima).

   The second major irony is inscribed in a report that had been written by the Tokyo Electric Power Company (TEPCO) three years
before the Fukushima disaster but was never delivered until March 7,
2011, four days prior to the tsunami that hit Fukushima. As NHK,
the Japanese national broadcasting service reported on October 3,
2011:

> Government documents show that the operator of the Fukush-
> ima Daiichi nuclear plant predicted in 2008 that a tsunami over
> 10 meters high could hit the plant, which was only designed to
> withstand a tsunami of 5.7 meters. But it failed to report this to
> the government until just before the March 11th disaster. . . .
> TEPCO had predicted that waves between 8.4 and 10.2 meters
> high could hit all 6 reactors at the plant in the event of an earth-
> quake similar to one that devastated the area in 1896. But the pre-
> diction was not conveyed to the government's nuclear safety
> agency until March 7th, just 4 days before the plant was crippled
> by tsunami. . . . TEPCO did not feel the need to take prompt
> action on the estimates, which were still tentative calculations in
> the research stage. But a Nuclear and Industrial Safety Agency
> official says it is regrettable that TEPCO did not start work on
> its tsunami measures right after it made the estimate 3 years ago.[7]

   The power of hindsight is often criticized for being just what it is:
the ability to make a judgment upon a situation after the facts are all
in. Obviously, it is easy to say now that TEPCO should have released
the report sooner. At the same time, it is clear that this is not the first
instance of not doing the right thing in the first place when it comes to
atomic power. We already have many examples in our "hindsight
cache." Unfortunately, in yet one more ironic moment, it takes a

disaster to get our attention. We can only hope that Fukushima's lessons are heeded all across this earth.

NOTES

1. E. Strickland, "24 Hours at Fukushima," *IEEE Spectrum*, November 2011, 35–42.

2. IAEA.org, "Radiation Safety," March 27, 2011.

3. CNN World, "TEPCO Hikes Radiation Limits," March 18, 2011.

4. H. Caldicott, *New York Times*, "Unsafe at Any Dose," April 30, 2011.

5. K. Belson, *New York Times*, "Japanese Find Radioactivity on Their Own," July 31, 1011.

6. H. Tabuchi, *New York Times*, "Japanese Tests," December 6, 2011.

7. CNN World, "TEPCO Hikes Radiation Limits," March 18, 2011.

Of Romance, *Soteigai*, Celebrations, and the Mundane

A keystone of the atomic mindset has always been awe, or what the Romantic poets and artists of the eighteenth and nineteenth century called "the sublime." The Romantic sublime is a concept that goes beyond the contemporary usage of the word as something highly pleasant and beautiful, as in, "That performance was just sublime!" It also encompasses the sense of terror and awe that can come from a moment in the sublime. Put simply, an experience with the sublime can take your breath away. In the atomic age, this phenomenon has been a strong thread that has woven the human imagination into quite interesting patterns. The early fascination with luminescent radium crystals certainly created an innocent awe as people became fascinated with the illuminating properties of this newfound element. The promise of radium, it seemed, was boundless. For dial painters and then workers in radioactive environments, however, the fascination turned toward the terrible—the terror—as they witnessed their own demise or the demise of others.

When the awe of the atom progressed to the advent of its most spectacular achievement—the bomb—the sublime moment reached its ultimate end. In the New Mexico desert, Trinity represented, at one and the same time, the awe-inspiring power of human ingenuity and "the destroyer of worlds." Now, there would be no turning back. Awe would gradually turn toward terror and fear, although some would still be held in the Romantic moment like the VIP observers on Enewetak Atoll in the frontispiece of this book.

And for the people of Fukushima, the romance with the atom became what the Japanese government and power company officials

Atomic Energy Commission Cowboy, circa 1955. Cattle were often used as subjects on bomb testing sites. (Photo provided by the Defense Threat Reduction Agency (DTRA).)

have referred to as *soteigai*, or that which is "outside the imagination." The argument used by purveyors of the term *soteigai* is that the meltdown of the Dai-ichi power plant was an inconceivable confluence of a natural disaster and a complex of technological and human breakdowns that could not have been predicted. In other words, *soteigai* is just another word for one of the essential excuses of the atomic mindset: "We just didn't know."

Celebrations go right along with the notion of romance, too, as the awe and inspiration wrought from the atom can make people get up and dance like, for example, the young woman dancing in the desert. She is dancing in the Southwestern desert to celebrate the first and only testing of an atomic cannon. Celebrations of the atom, as we have

Cannon Dancer. Woman performs a pas de deux at the Upshot-Knothole Dixie event, April 6, 1953. This was a celebration in honor of the first and only test of an atomic cannon. (Photo provided by the Defense Threat Reduction Agency (DTRA).)

seen, can also be used to heal the pain of the atomic age, such as with the Bikini Islanders and the Navajo. In their cases, the celebration is of the human spirit and humans' ability to celebrate life even in the most frightening and devastating of circumstances.

Finally, the atom is found in the mundane experiences of everyday life. The "AEC cowboy" provides a wonderful image of the fact that people, in a wide variety of ways, were involved with the atom to simply make a living and just plain get the job done. The cowboy had the mundane task of branding cattle that had been grazing on or near atomic test sites to warn anyone who might be interested in filling their freezer. He also provides a most fitting symbol for the end of our stories of the atom, the metaphor of "brand." The cowboy is branding cattle, just as we have branded the Earth with intensive amounts of radioactive matter. We have made our mark across the globe with a radioactive brand that will be on Earth for countless generations of fauna and flora to come.

APPENDIX

Living in an Atomic World: Voices from "On the Ground" of the Nuclear Age

These three interviews/conversations bring four contemporary voices to the atomic romance so that we can hear what it is like to live with and work within different facets of the atomic romance. The interviews took place in 2010 and were conducted by Robert and Evelyn Johnson.

The first interview is with Jack Niedenthal, who has been the Trust Liaison for the people of Bikini for the past 25 years. He is a main conduit between the Bikinian people and the U.S. government and thus plays a very important role in communicating with legislators and other government officials who make decisions about funding and related support for the people of Bikini. In addition, he is a voice for the Bikinians to the greater world and an advocate in their struggle to recover their precious land.

The second interview, with Dr. Michael E. Mullins, came about as a result of an informal conversation I had with him (we both work at Michigan Technological University) about this book. He told me that he had been one of the early environmental scientists in the 1970s at Oak Ridge, Tennessee. He only needed to tell me a few brief stories of what it was like to work there before I asked him if he would be willing to share them with a wider audience. He graciously agreed. The results are some fascinating narratives of working within the bowels of this enormous atomic weapons development complex as a not-always-welcomed environmental engineer there to oversee cleanup of radioactive wastes, among other controversial things.

The final interview is with two women from Oxford, Ohio—Linda Musmeci Kimball and Yero Peterson—who played central roles in fostering the cleanup of the Alba Craft uranium milling site in the 1990s. Their voices have already been heard in Chapters 7 and 8, but here they tell some candid stories of the concentrated effort involved in advocating for cleanup. In addition, they argue that there is a great need to be advocates for education about the atomic age, especially for young people who have grown up without perhaps much background in this fascinating, terrifying, and ongoing history that will continue to affect future generations.

A CONVERSATION WITH JACK NIEDENTHAL, TELEPHONE INTERVIEW, JUNE 18, 2010

RRJ: You are the Trust Liaison for the People of Bikini. Is that your official title?

JN: Correct.

RRJ: How did that role come about for you and for the Bikinian people?

JN: I was in the Peace Corps for three years from 1981 to 1984 on Namu Atoll. I was just about to leave the Marshall Islands to go to graduate school at Penn State to get a Ph.D. in economics; I had made all the arrangements, told my parents, etc. One day I was sitting in a restaurant in Majuro, the capital city of the Marshall Islands, talking to a friend from Namu in Marshallese. Namu is a very isolated atoll. I was out there for three years, so I learned the language really well. As I was talking, some Bikinian men overheard me speaking in Marshallese—they were very impressed. They told me about a teaching position on their island. The Bikinians were thinking about moving to Maui, at the time, en masse. And I thought, well, I already had spent three years on the outer islands; I can do one more year on Kili Island. That quickly turned into three years. I say "quickly" because in the blink of an eye, there was so much activity going on down there.

The Bikinians had just gotten their $20 million trust fund from the U.S. government to help them survive. I was teaching school and working with the Council creating sports programs for the community. The Council also asked my advice on various issues from time to time, so I would usually attend their meetings. They had a lot of social problems because Kili was such a small island. For example, at the meetings they would complain about how their kids were playing in the streets every night; they felt they had do something about this. I suggested, "Why don't you build a recreation center?" Suddenly, within a few days, there was a bulldozer moving earth to build the recreation center. For a young guy this was a pretty heady experience. It was actually a lot of fun. We built tennis courts, a basketball court and made a softball field. We did all kinds of great projects.

And that's what I was doing when my predecessor—Ralph Waltz—suddenly collapsed one day in 1986 over Christmas vacation with brain cancer. He was completely incoherent; he couldn't talk. It saddened me because he was a friend of mine. My only instruction at the time, from the mayor and the Council was, "Just go do what Ralph was doing"—which was not much guidance. I went to our office: All the filing cabinets were locked and there were no keys, there were no discs for the computer (this was before hard drives), I had no idea where anything was and no one could tell me where things were. So I pretty much had to start that job from scratch. This was late 1986. I have been doing the same job ever since.

RRJ: How is that position supported, institutionally or financially?

JN: Institutionally, when the Bikinians got their trust fund in 1982, they needed a Trust Liaison, someone who would act as a go-between for the Bikinians and the U.S. government. The U.S. government has ministerial oversight over the Bikinians' Resettlement Trust Fund. At that time, 1982, the trust fund amounted to $20M. Subsequently, that trust fund, after I came on board, was given another $90M by the U.S. government. So it became $110M, and every year there is a need to produce and follow an annual budget. I am hired by the Bikinians and paid by the trust fund. It's the Bikinians' money, but as I said, the Interior Department has the ministerial role. They make sure we're spending the money responsibly. The trust fund has two competing functions: one use for the funds is to take care of Bikinians where they are living now, and the other is to clean Bikini Atoll.

My role is not just administering this trust fund on behalf of the Bikinian people. I am also responsible for working with the media to make sure they get the story of the Bikinians told in an accurate way. The media has always been important to the islanders. My marching orders have always been, "Make sure our story continually gets told to the world in any way you can get it out there."

RRJ: In your book, you have a listing of the demographics of where the different Bikinian people live now. Is that still pretty accurate from your book?

JN: If you go to the "Bikini Facts" part of our website, there is probably a more updated version. The population increases between 4 and 5 percent each year. Right now, in 2010, there are about 4,500 Bikinians.

RRJ: It has been said that the atomic tests in the Marshall Islands in the 1940s and 1950s were not only to test the weapons and study the levels of radiation, but there are claims that there were explicit, consciously designed tests performed during these events to study the effects of radiation on the Bikinian people without their knowledge. Is this accurate? Was there conscious, premeditated, uninformed testing?

JN: When the US declassified the testing documents during the Clinton administration, we discovered that there were a number of scientists making some remarkably candid comments. One scientist involved in the testing program said something like, "studying the Marshallese after the nuclear tests was better than studying rats." Obviously, that's a crude statement, but I believe he said it thinking that no one would ever see the documents.

My opinion is—and again I am an American with Marshallese citizenship, my wife is a Bikinian and I have Bikinian children, so, I do have some bias here—whether or not irradiating humans was done intentionally, the end result was that they have studied the people involved in the nuclear testing out here. And you can understand why that leads some individuals to speculate as to whether or not this was done on purpose.

When you read the facts from the day in 1954 that the Bravo hydrogen bomb was detonated, they tell a sordid tale. The US ordered all U.S. military personnel below decks, they were told to close all hatches, doors, windows, etc., because they knew the test was going to be a powerful one. At the time the U.S. military was sending out these orders to their own men, they also knew they had all these Marshallese people living on outer islands that were clearly in harm's way. They never said a word, no warnings, not even a hint as to what was about to happen. The Marshallese people in the Northern Marshalls experienced two suns rising that day, one in the east as usual, and also one in the west. Then it began to snow. It was a very surreal event for them. As I said, they had no warning, no one told them this was going to happen. Even though the Bikinians were far away from Bikini during these tests, they moved back in the 1970s and lived there for seven or eight years, not understanding that they were ingesting cesium-137. The end result was that the people of Bikini were regularly studied for seven or eight years after they were moved off in 1979. The cesium-137 had gone up into the food chain, into the coconuts and into anything else that was growing on the land. At the time, it was the largest known population of people that had ingested this type of radiation. So the U.S. scientists studied these people, and they still do study the other Marshall Islanders who were living in the Northern Marshalls at the time of the testing. They have a program set up where these people can go to special doctors to be studied and taken care of if they get cancer.

The end result is what one has to look at. There's a lot of evidence out there that suggests that these people were set up to be studied like guinea pigs. But there is also evidence that suggests that the U.S. military and their scientists—for a lack of a better way of saying it—didn't really know what they were doing; at times they appeared to be just boys playing with toys. In the end, however, they used their own mistakes to study the Marshallese people affected by the nuclear testing. That's kind of where I am on this, I see a lot of humans making mistakes, but I don't argue with people here when they say they were used as human guinea pigs. I can easily see how people could come to that conclusion.

RRJ: This theme of "guinea pig" comes up all the time. And there clearly were people who were tested in various places around the world, without their knowledge.

JN: And that is exactly what happened out here—you have indirect evidence out there that people were actually used for studies. So people just connect dot A with dot B, without really giving it much thought. And it's logical to do that. I think the way the Atomic Energy Commission and the Department of Energy went about their work out here, especially in the early days after the testing, was flawed. There were so many times where U.S. officials misled people and even lied to them, and other times where they made some honest mistakes but just kept trying to explain them away in a fundamentally dishonest way. The way they handled the situation out here lent itself to all kinds of speculation, theories, and even chaos when it came to getting to the truth of what really happened out here and what the true radiological situation is even today. And that, to me, is the biggest problem in the Marshall Islands today when it comes to the nuclear testing: The psychological health of the victims at times seems to outweigh the physical harm done to them from the testing.

RRJ: In terms of that psychological or physical health, what kinds of problems linger there for the Bikinian people?

JN: The Bikinians and other people on the Marshall Islands suffer from many different kinds of cancers. The medical issue that I see that is most prevalent with people here is thyroid cancer. And you see these kinds of cancers everywhere out here. My father-in-law died of a stroke, but he had precancerous thyroid nodules. My wife's uncle died of cancer of the thyroid. The people who actually got "snowed on" [after Bravo] on Rongelap and Utrik Atolls have all kinds of thyroid issues. That's what jumps out at you.

But there are other issues. My mother-in-law died of cancer of the uterus at the age of 49. Every family has these stories: cancer deaths at an early age. But you also have a lot of people with poor diets, which leads to diabetes, high blood pressure, and obesity, so there are other health factors at work in the Marshall Islands. And you have people with financial issues and the stress that that brings on families. Ironically, the nuclear victims were compensated with Western dollars, and therefore you've got a lot of Western types of money problems that you experience on a daily basis.

In addition, you have a victim mentality that's really problematic if you are trying to raise your children to be motivated in life. At my office, on any day of the week, you will find 20 or 30 people sitting on our front porch doing nothing except waiting for the next payment. They tend to blame the United States for everything. That is the victim mentality. And that's very prevalent. I find it very depressing to deal with on a daily basis. Not everyone is like that, but I would say the vast majority are. This whole series of events has created a welfare state. But it's even worse than that, because there are no jobs in the Marshall Islands, so even if you wanted to go out and work, the offerings are slim. There's a 40 percent unemployment rate in the Marshall Islands.

RRJ: Are there controversies between the Bikinians and others who live in the Marshalls? Is there resentment?

JN: No, because there were three other atolls that were compensated for the nuclear testing, and also anyone in the Marshalls who has had particular types of cancers have also been compensated. The Kwajalein Missile Range operated by the U.S. government in the Marshall Islands pays out millions of dollars every year to the Marshallese landowners, far more than any of the Bikinians get. And I think you have to understand who the Marshallese people are. There are all kinds of Marshallese people who have married into our community who are not Bikinians. I am one of them. As soon as you marry, you become part of the community and you get an extra $800 per year, maybe. But it's $800 more than you would get otherwise. And there's an unbelievable amount of sharing that goes on in the Marshall Islands; it is their custom. So while there may be a little bit of resentment, Marshallese people are generally not like that. They believe in destiny and would say that "My destiny is that I don't get a compensation payment, but here's this guy who does, and I'm not necessarily jealous of that person." That's just the way it is.

RRJ: Are there still food and starvation issues for the Bikinian people? And where does their food come from?

JN: Two years ago, we had a $10M budget [FY 2008]. With that, we were able to have scholarships; we had a medical plan for people living in the United States. We have over 500 kids in the United States attending schools from grade K to college. Then the stock market crashed. Just like most endowments in the United States, our budget is driven by how well the U.S. stock market does. Just like everyone else, we lost a huge amount of money in 2008. It was so bad that the U.S. government, exercising their ministerial role via the U.S. Interior Department, said in 2009, "You need to cut your budget to $7.5M." And then, this year [2010], they came in and said, "Now you have to cut it to $5M."

So, in the space of two years, we cut our budget by 50 percent. That eliminated scholarships for our students; it eliminated our medical plan and numerous other programs we had funded with our trust. And it also eliminated a food plan we had for people living on Kili where we would spend about $250,000 a year supplementing their diets. We quickly ran into a situation where we didn't have enough food on Kili. We went to the Interior Department and told them that the budget reduction was having a serious impact on our lives, that we needed food for the people on Kili. They were sympathetic and did allow us to add on some money so we could buy food.

There were other problems caused by these drastic budget cuts. Normally, we had 24 hours of power produced by the power plant on Kili Island. We had to cut back to 10 hours per day, which is not enough to keep food refrigerated. So, because of the budget cuts, the Bikinians suddenly had to take a huge step backward in their standard of living. Meanwhile, even though most Bikinians aren't that well educated when it comes to the U.S. financial markets, they watched on CNN as the U.S. government bailed out all the banks, some of which had a major role in creating the problems in the first place. They would ask me, as the American who has been tasked to explain the outside world to them, "Hey, Jack, we did nothing wrong. It was the U.S. stock market that crashed, and the banks created most of the problems that caused these markets to crash. We see these car companies that don't know how to run their businesses, they're getting bailed out with billions of dollars. We did nothing wrong, yet we have cut our budget by 50 percent. We look at the U.S. government that is trillions of dollars in debt, yet they don't seem to care about cutting their budget. Why is this?"

So the Bikinians look at what has happened over the past couple of years and say that it is not fair. And, of course, it's not. And, I don't even know what to say to them when they say, "How come we don't get a bailout? They're bailing out all these businesses in the United States, yet we get nothing?" It is hard to explain these issues.

RRJ: In your book's interviews, some of the Bikinian elders, and maybe not just the elders, refer to Kili as a prison. Does it still seem that way?

JN: I lived there for three years from 1984 until the end of 1986. Sometimes, people would go down there to visit, and they would see palm trees and a sandy beach, nice little houses along the edge of the island, and they would think it very beautiful; but having lived there for so long, everything quickly turns in on you. Everybody is living on top of everybody else. It's crowded. There are some things to do there—there is fishing, but it's only good for six months out of the year. People there are almost totally dependent on outside food sources. Because of my job, I have to travel to Kili once or twice a year, but I don't even like doing that. The people don't bother me; it's the island itself. Though I will say that how people treat each other there is entirely different than how they treat people when they are living somewhere they like. And what's happened over the years, people have moved away from Kili and the Marshall Islands.

We have over 1,000 people living in the United States: in Arkansas, Oregon, Costa Mesa, and other U.S. cities, we have people who have decided, "I am not living this kind of life in the Marshall Islands anymore where I am lucky if I can get $2 an hour. Instead, I am going to Arkansas and work for Tyson Chickens where I can step off the plane and make $7 or $8 an hour and have my children attend nice middle-class schools in Arkansas so they can get a better education." There are ways to get off of Kili, and people have taken advantage of that. If you have children, and they are bright, you get them out of the Marshall Islands to Arkansas or Oregon, or somewhere, to go to school.

RRJ: That leads to another question I have about the state of education in the Marshalls. You mentioned that you have lost a lot of money because of the stock market, but can people receive a decent education there?

JN: I am heavily involved in education in the Marshall Islands. I am president of the best private school in the Marshalls, the Majuro Cooperative School. Several Bikinians attend this school. It's a Western Association of Schools and Colleges (United States) accredited school. When the accrediting association comes to visit us, they always comment about how fantastic our school is. They're just amazed. We hire American teachers to work in our school; it's almost like being a volunteer because it doesn't pay much, but we do get certified teachers to work for us. I am also a regent at the College of the Marshall Islands, the only college in the Marshall Islands. We have four to six Dartmouth graduates who come out to teach in the elementary schools on Kili and Ejit Islands where most of our people live, and they have been coming every year for the past decade.

The Bikinians made a conscious decision, almost as soon as they got their trust funds, that if they were going to be nomads wandering around from place to place because Bikini was too irradiated to live on, then education would be essential to better the lives of their people. And we have probably gone from being, at that time, the most uneducated people on the Marshall Islands (the Bikinians were referred to almost as being hillbillies or country bumpkins) to the most educated group of people in the Marshall Islands. We have emphasized education for as long as I have been here and worked for them—for 25 years.

Even though we made many sacrifices when we cut our budgets over the past two years, we did not touch the funds set aside for the Dartmouth Volunteer Program, which costs us about $75,000 a year to operate. So, in a sense, to answer your question, I think education is still as important as it ever was, and probably even more so. As our budget goes down, people realize that the only way they are going to get ahead is through education.

RRJ: I am also writing a chapter on the Navajo experience in the southwest—the 1979 Church Rock Disaster. It was a huge flood of radioactive materials that came out of a tailings dam that is pointed to as the largest nuclear accident in the United States, with the exception of bombs testing. The Navajos, in dealing with this problem (and it has now been over 30 years), is that every 10 years they hold a festival that celebrates their spirit and their heritage. And they try to turn to the other side of their problem. It seems, in some ways, that the Bikinian people have a similar thing in regard to promoting tourism, and now I have read your story in the *Bulletin of Atomic Sciences* about the Heritage Site designation. Do they have a spirit like that, in some ways? Do they want to celebrate their heritage?

JN: Every year, and this goes back to the 1970s, they have a commemoration called Bikini Day where they have speeches and support many community-wide activities. We invite the Diplomatic Corps of the Marshall Islands, the American, Japanese, and Taiwanese embassies, and other dignitaries in the Marshalls. We have this event on Kili, which appears, on the face of it, similar to what you're describing with the Navajo people. About 10 years ago, the national government in the Marshall Islands made Bikini Day into a national holiday to be commemorated every March 1st. Now, instead of calling it Bikini Day, it is called Nuclear Victims Day.

In terms of spirit, one has to understand that for Bikinians and other Marshallese people, land has a different meaning than it does in Western society. Land is not a commodity that can be bought and sold; it belongs to the clan forever. My wife and my children have land that they are attached to that will be theirs forever, and in turn this land will be passed on to their children. They believe their land is a gift from God. So when your land—your gift from God—is taken away from you and you have to live on other people's lands, spiritually, especially for the nuclear victims of the Marshall Islands, it's crushing and sad. All of our elders are being buried on places that are not their own, and every time there is a funeral (I attend them all), they make speeches about how disgraceful it is that they're not being buried on their own land.

RRJ: Over the years, there have been so many stories and documentaries and television specials about the Bikinian experience. Some of them are so incredibly powerful. For instance, in one where the U.S. military staged the 1946 request of the people to move, they restaged it and refilmed it a dozen times in a Hollywood outtake way. And you see this absurdity. But they are all such powerful stories. I am wondering, have these had lasting effects? Are there particular works or things that either you or perhaps the Bikinian people know of or point to that this has had an effect on the rest of the world, or maybe on how this has been their fate?

JN: The one thing that I am happy about with what I have done with my life is writing my book [*For the Good of Mankind*]. Most of the people in that book are now dead, and I had the opportunity to spend time with them on Kili, hour after hour, day upon day, for several years in a row. And even after I went to work in my office in Majuro for the Bikinians, these old men all knew they could come into my office and just talk the day away. I encourage that by always having some coffee to serve. Even though they don't necessarily come to ask for my help or to ask me to do something, I have always enjoyed the visits from the elders. They're funny, they're more laid back, and they have a nice perspective on life. Many of them have this powerful attitude that they have actually done something for mankind. But some of the older people also have other perspectives. For instance, Kilon Bauno was one elder I interviewed a lot. He was very bitter about the entire affair, the exodus, the neglect, etc., at the hands of the Americans. He was angry. We toured with one of the films that was made about the Bikinians, *Radio Bikini*. If an American in the audience asked him a question about his experiences, he would treat that American as if he or she had set the bombs off.

RRJ: Did he [Kilon] go to the Academy Awards with you?

JN: Yes. And he was very effective as a speaker and as an actor. I believe he did a lot towards helping the Bikinian people get compensated because he was genuinely angry and bitter about all the lies and deceptions that the American government used against the Bikinians. I enjoyed being around him; he was a charming man.

Another person I treasured being around was Lore Kessibuki. He wrote the Bikinian anthem. He was a poet and always had a philosophical take on everything. He couldn't speak a word of English. My greatest accomplishment, which turned into my own gift, was learning the Marshallese language. Having the ability to understand the Bikinian elders when they tell the stories of their exodus has been so important to me. Lore would sit with me by the beach or in his house and just talk for hours; it was better than having a television set. He was softspoken and had a beautiful personality. He sincerely believed he had done something for mankind and that that was something the Bikinians should be proud of. He was never bitter, although he could get a little angry when talking about starving after they were moved off Bikini to Rongerik Atoll. That's when the Bikinian people were actually suffering the worst. But most of the time, he was introspective and philosophical. He was that kind of person. His views became mine. The Bikinians did do something outstanding for mankind. One way to look at it is that the Cold War was fought and won on the shores of the Bikini Atoll. I think a lot of Bikinians look at it that way; but, as I said, there are some who are still bitter about what the Americans did to them and they way they perceive they are treated now.

The Bikinians always understood the power of the media, why it's important . . . and that's one big reason why I have my job. Sometimes media people don't understand why I do what I do. I am an American: Don't the Bikinians despise me for has happened to them? But the Bikinians are the ones who always make sure that when the media comes around, I am there to act as a buffer, as a translator, because they want to make sure that the story gets to the person from the media. In a general way, the people of the Marshall Islands are so shy. With many of our elders, without some kind of conduit or catalyst to get a discussion going, the media would never have heard their story. But, conversely, when you show them a *New York Times* magazine article or television show about the Bikinians, and say, "Isn't this great!"—it doesn't matter to them. I don't know if that's the right way to express it, but it doesn't mean as much if, say, you were on ABC Television, what it would mean to you. To most of our elders, being in the media doesn't mean anything. It's not a big deal. And it's something I've recently discovered.

This is an interesting example: I buy and sell things over the Internet. I bought 1,000 copies of *Forbidden Paradise* that tells the whole story of Bikini on the Discovery Channel. I've been selling those for about 10 years, and still have about 400 left. Then, I wrote, directed and produced two feature films that were released in 2009 and 2010; they're amateur films in Marshallese with English subtitles. The films have nothing to do with nuclear testing but are two modern-day stories about the Marshall Islands culture. Each film sold about 2,500 copies in two weeks.

RRJ: Why is that?

JN: Imagine growing up, all your life, never seeing a feature film in your own language, set in your own country, talking about things that are relevant to your own culture, never seeing that as a kid. And then, suddenly, here are these films. But when you see the documentaries concerning the nuclear issues, well, everybody knows about these issues. It's a depressing piece of history, and most people here have seen it over and over again; the story is known.

RRJ: Sometimes, with indigenous cultures around the world, they can be quite sensitive about how they're represented. They can be insulted, angry. They can resent having white people there. I wonder if there are any stories or tales depicting the Bikinians in ways that they find more acceptable than others. Do they have this kind of feeling?

JN: One story jumps immediately into my mind. In 1994, the *New York Times* magazine did an article, a cover story. There was a young journalist who at the time had never done anything big. He had been in Las Vegas at the same time we were there for a meeting, and his intention was to do an article for *Outside Magazine* about the tourism program we were considering starting on Bikini. At that time, it wasn't yet a reality. One of the older Bikinians came out of the meeting, and when asked by the journalist what our plans were for Bikini, responded by saying, "Oh yeah, we're going to do the Dive program here to promote scuba diving and tourism, nuclear waste over here on this island, and we're going to launch missiles from this part of the island." The journalist just looked at this guy and said "What?" Intrigued, he immediately called the *New York Times* and said, "I have this great story." They said they would put it in the weekly *New York Times* magazine.

For him, as a young journalist, this was a dream come true. Nuclear waste was part of the issue, at that point. So, he wrote the story, and immediately it spun out of his control; the editors took it over. He called me to apologize, saying that he didn't have much control over the story, that the editors had changed some of what he wrote, that this was a big career opportunity for him and there was nothing he could do about it.

In the article, the *Times* used a picture of King Juda and his family in 1946. Probably the most prominent person in my life has been Tomaki Juda, King Juda's youngest son, who is now our senator. In the photo, Tomaki, as a five-year-old boy, is standing with his family, stark naked. I knew this photo, if published, would infuriate him.

The *New York Times* had faxed me the picture and asked me if I could identify the people in the photo. I can remember, in the 1980s, in the film *Radio Bikini*, there is a little boy running around naked. That was also Tomaki. Robert Stone, who produced the film, came down to Kili at one point to show some of the footage he had discovered. He showed many outtakes— where they do Cut 1, Cut 2, etc.—and everybody laughed at that as the people always seemed to be saying the same thing. But that was Tomaki's father, and the laughter at the scenes that had his father in them embarrassed and angered him. And then, all of a sudden they show this clip of Tomaki running around naked. Everyone knew who it was, and everyone laughed. Tomaki got up, stormed toward the projector, and shut off the film.

So when I saw this *New York Times* picture, knowing how Tomaki would feel about it, I wrote the editors and begged them not to publish it. They said they needed it to show the innocence of the people at the time they were moved from their islands. I again urged them not to use it. Well, this was the *New York Times* and I am just Jack Niedenthal, so they went ahead and published it anyway.

Once the article came out with the photo, I decided that I wasn't going to show it to Tomaki. One day, forgetting that I had the article on my desk, Tomaki came to my office. To my horror, he picked up the magazine and started thumbing through it. I knew, immediately, there was going to be a problem. The magazine had Bikini on its cover, so he knew there had to be something in the magazine concerning his people. Suddenly, he saw the picture of his family. He stood up, angrily took that *New York Times* magazine, which was probably 1/4" thick, and jammed it into my paper shredder. He growled at me, "Don't ever show this story to anybody!"

That illustrates the point that while the Bikinian people understand the power of the media and what it has done for them, they also know that working with the media always has a potential downside.

RRJ: That's amazing—because you would think that with a newspaper like the *New York Times*—these are highly educated people who, you would have hoped, had had some education or cultural sensitivity that might have made them think twice about it.

JN: This was back in the days of faxes. I wrote and faxed volumes of pages to try to get them to change their minds regarding that photo because I knew it would devastate someone I really cared about.

EVJ: So how did you work it out with Tomaki? Did he forgive you? Did he blame you?

JN: You have to understand. Tomaki is like an older brother, or uncle, to me. I had worked with him for 10 years at that point. We were and are very close. There are times when he will come into my office, lock all the doors, and talk to me about personal or business issues that I know he discusses with no one else. Yes, when he saw that photo he was outraged, and although he didn't really blame me—I showed him everything I had written to ask the *NY Times* to keep the picture out of the story—it made him deeply mistrust the media. He now rarely does interviews. And I think that photo is a big reason why.

RRJ: On the other side of these insulting stories, are there some stories that the Bikinians hold up as exemplary about them that they would like others to see?

JN: Yes, I would say that the Bikinians love 95 percent of what has been written about them. In some ways, the media saved their lives. And that's the way they look at it. No one was listening to them in the U.S. Congress. They had no traction, whatsoever, in the United States in terms of trying to get their situation resolved. The only thing that saved them in the early days was *National Geographic Magazine*, which started writing about their plight in 1946. The *New York Times*, and later on, *Time*—virtually every major news publication, whether it was magazines, newspapers, television—has at one time or another produced their own version of the Bikinian story.

I have probably dealt with every major news outlet in the world—not an exaggeration—and the Bikinians love this because nothing sticks a pin into a Congressmen and gets their immediate attention as when a story gets play in the media. That's when their own constituents start saying, "Hey, what about this?" In fact, we just had a hearing in May, in Washington; and the whole thing came about because of a *Newsday* article in the fall of 2009. That immediately prompted a Congressman to want to have a hearing.

RRJ: Is the Supreme Court case concerning the future of more settlement awards now a "done deal," that there will be no future hearings in the courts to award more monies?

JN: The Supreme Court rejected the Bikinians' case. Now we can go to Congress and say, "You set up this tribunal in the Marshall Islands, but you didn't give them enough money." Congress can refer the case to Federal Claims Court and ask the court to review it and tell us if this number that the tribunal came up with is an accurate number or not, in terms of what the should be the level of compensation above and beyond what they have already received. So that's what we're pushing now. And there's some traction for that. There are some Congressmen who are saying, "Yes we did this to the people of the Marshall Islands, the situation is still unresolved, so why aren't we trying to make these people whole?"

RRJ: The Bikini World Heritage Site. Is that something that is moving along? And tied to that, in a more general way, are there things people can do?

JN: Bikini Atoll became a World Heritage site in August of 2010. The Bikinians believe this to be an important event in their history because this powerful organization has acknowledged their contribution to the world.

In answer to what people can do, I just tell people to ring up their Congressmen, or take a few minutes to write a letter, and ask why this has situation gone unresolved for so long. What's going on in the Gulf of Mexico [the BP oil rig explosion and spill], right now, is very relevant. Are you aware of the Marshall Islands connection to that . . . That the oil rig flew under the flag of the Marshall Islands? Someone asked me what I think about this, and I made a very flip remark, which I wish I could take back; but I said, "OK, now we're even." Here are these people saying how devastating the oil spill is, and they're trying to blame the Marshall Islands for this big environmental disaster. We've been waiting 67 years to try to get our situation rectified—cry me a river.

RRJ: Did you see the *Colbert Report* episode on the Gulf spill and the Marshall Islands?

JN: Yes. Stephen Colbert declared war on the Marshall Islands. Marshallese are really offended by that. Oftentimes, during the Iraq War, when people like Michael Moore were making fun of "the only countries that are supporting the Iraq War are these goofy countries out in the Pacific, like the Marshall Islands"—What they don't understand is that we have men and women in American uniform, actually there, fighting in Iraq.

EVJ: Do you think that, had the Bikinian people not been converted to Christianity, they may not have agreed with the U.S. generals who were trying to convince them that they were serving God and mankind? That spiritual system seems to set them up.

JN: People often talk about how it was very manipulative for the American officials to ask the Bikinians on a Sunday, after church, to move from their islands because it would be for the good of mankind, that they would be like the children of Israel, etc. On the Bikini flag it states in bold black letters: "Everything is in the hands of God"—I don't think there is a flag on earth that says anything like that. In the film clips, even though U.S. Commodore Wyatt asks King Juda over and over again in many different ways if the Bikinians would be willing to leave Bikini, Juda's answer was always the same: "Everything is in the hands of God." I think Christianity overlaid itself on the culture of the Marshall Islands. The two films I have done are all about this shaman woman who is an evil witch, but she is still reading the Bible every day. Even though Christianity has overlaid itself, the culture is still here in full force. So, in answer to your very interesting question, I don't think it would have made any difference. The Bikinians still believed in a god before the missionaries came.

RRJ: Is it still true that the United States still lobs a few Minuteman missiles over you every month?

JN: Oh yeah. Look, as an American, I have no fear of any other country in the world. I've seen these missiles hit a dime after they have been shot from Vandenburg Air Force Base in California all the way out to the Marshall Islands. They're so accurate.

A CONVERSATION WITH DR. MICHAEL E. MULLINS, PROFESSOR OF CHEMICAL ENGINEERING, MICHIGAN TECHNOLOGICAL UNIVERSITY—DECEMBER 16, 2010

RRJ: Tell me what your job was at Oak Ridge. What was it like to be an environmental scientist/engineer at Oak Ridge in the Seventies?

MM: I graduated from Georgia Tech with a master's degree in chemical engineering in 1976. I went to work at Oak Ridge at the gaseous diffusion plant, the K25 plant, where they did uranium enrichment. I worked there for a couple of years, went to MIT, and came back. I worked at all three sites: K25, Y12 (Weapons Plant), and X10 (Laboratories). I worked as an environmental engineer. We were so new that we were housed in a temporary classroom building. This was not too long after the EPA was formed in the early '70s. The new solid waste rules (RICRA) came into effect just as I started. So we were faced with all of these new solid waste rules and how to abide by them. It was a bit like the Wild West. The environmental group grew to several hundred people. At that time—1977 to 1978—we only had six or seven and were mostly putting out fires.

RRJ: At that time, the Oak Ridge main building was the largest single-story building in the world and employed tens of thousands of people.

MM: That's true. The Oak Ridge facility complex had about 250,000 people (workers and their families) during World War II.

RRJ: Can you describe some of the activity at Oak Ridge that made it necessary for environmental scientists to be on site?

MM: Over the many years of weapons development, they had created all sorts of wastes, including holding ponds and quarries. The holding ponds were the receptacles for all the radioactive sludge, and the quarries were supposed to be the recipients for the sanitary waste. I started doing inventories of all the stuff that had accumulated since World War II. The K25 plant was built during World War II as part of the Manhattan Project to enrich uranium by the gaseous diffusion of uranium.

The gaseous diffusion process is messy. We took yellow cake (mixed uranium oxide) and processed it to make it a gas (UF_6). We treated it with hydrofluoric acid (HF). There was a tank that contained tens of thousands of gallons of anhydrous HF. On a hot summer day, the temperature was high enough to flash evaporate the acid. One drop of it can dissolve an arm, so there was enough to kill everybody in Oak Ridge and maybe Knoxville if there had been a significant incident.

HF is used a lot because you can dissolve anything in it. Even when mixed with water, it can dissolve glass. We had a huge amount of HF. When HF reacts with yellow cake, it forms UF_4. When UF_4 reacts with fluorine, it becomes UF_6. We heated it and sent that through membrane separators (there were over 100), which made the uranium-235 portion in the gas more and more concentrated. They were decommissioning some of the membrane separators when I got there.

K25 used as much electricity, at that time, as New York City. The TVA had three coal-fired electric plants and two immense hydroelectric dams in the Oak Ridge area. We had a lot of power going into K25. Various parts of the process for making UF_6 produced sludges, potentially nasty stuff. The gaseous diffusion plant also made gas emissions.

The attitude at that time was that anything concerning national security and the DOE trumps anything the EPA wants to know. Back in the day (the 1940s–1960s), a lot of things hadn't been well disposed of. For instance, we found hot spots deep in the woods where radioactive materials had been dumped. They recently closed down a holding pond I was inventorying. We dug up the old sludge from the pond and put it in several thousand 55-gallon drums. The old pond was not lined, but they eventually built a new one with a liner. That old pond had been in operation from 1945 to 1976, approximately, and inventorying exactly what was in the sludge was a problem. The inventory showed we had tens of thousands of pounds of uranium unaccounted for. Where did it go? Did it seep through? We couldn't find it. Were the inventories off? Maybe it went into the Clinch River.

In the old K25 building, which had one of the largest floor spaces in the world, they marked the floor with a grid. At each intersection, they would put a 5-gallon metal bucket filled with sludge or other waste. The buckets needed to be far enough apart so as not to create a "criticality incident." In criticality incidents, fission occurs where neutrons are released near fissionable materials. Long before I got there, a few people were killed and injured during criticality incidents.

RRJ: What was the process for filling the drums?

MM: They would drain the pond and fill the drums with a backhoe.

RRJ: Any other incidents you can remember?

MM: At K25, in the '70s, they were clearing the old cascade, where the enrichment facility was built. They found hundreds of gas cylinders—and only some were labeled and some were rusted. In one of the ponds, they built a platform to hold the cylinders in the middle of the pond. They got a guy with a high-powered rifle who, from a safe distance, would shoot a hole into the cylinders and drop them into the pond. This is absolutely true.

Another thing . . . they had started to segregate the waste. There was classified waste, radioactive waste, and toxic waste—sometimes all three. We were one of the largest users of polychlorinated biphenyls (PCB's) in the world. We had hundreds of thousands of gallons, mostly in electrical transformers. We also had radioactive PCBs. We even had classified radioactive PCBs. We had to figure out a way to get rid of them.

I remember, clearly, going to a plant managers' meeting, and they said, "Well, you know it's getting harder to get these PCBs for our transformers. We think the EPA is going to ban these PCBs, so we need to figure out what to do." I said, "They have replacement silicone transformer oils that are made of silicone fluids, so we use those instead." But you can't backfill the old transformers because the old transformers are contaminated with PCBs. The plant manager's solution was to buy up the world's supply of PCBs, and I said, "That's a really stupid idea." I was put on the night shift for two months after saying that.

K25 was going to be the recipient for all the PCBs for the entire nuclear industry: Hansford, Portsmouth, Ohio, upstate New York, what have you, so we proceeded to design a nice storage building. After we had just finished it, I got a call at 2 a.m., at my house, and was told that I needed to be down at the storage facility, right away. So I hopped in my car. There I encountered lots of snow-covered trucks (this was in Tennessee, where there was no snow), filled with barrels. We were told that we needed to put the barrels into the facility and that they couldn't tell us where they were from. The DOE manager said, "If you mention this to anyone, it's your job." That was the culture there. I had been there only nine months and had been threatened already.

They had a dozen semi trucks full of rusty old 55-gallon drums that were leaking who knows what and were full of rags and dirt. So we loaded them into our new facility, without a word. And the guy from the DOE said, "I hope you realize this is all secret. You can't talk to anyone about this." The only reason it was secret was that whenever the DOE screwed up, they didn't want the public to find out. So, whenever they didn't want the public to find out, especially environmentally, they labeled it "classified."

My main experience, to summarize one thing I learned in the environmental group at Oak Ridge, is that they used secret and top-secret classifications on environmental issues whenever they didn't want the public to find out, to protect their jobs. This was a specific case where I was threatened if I ever breathed a word about it. We took the 55-gallon drums, bought 65-gallon drums to put them in, and sealed them up. Finally, in the 1990s, they finished building an incinerator for all the DOE sites to get rid of all their radioactive and classified PCBs. I understand that hundreds of thousands of gallons of PCBs were destroyed. What I started designing in 1977 didn't go into operation until 15 years later.

MM: Can I tell you another story?

RRJ: Please do.

MM: They had secretly converted our old powerhouse at Oak Ridge—which was no longer a powerhouse—into a biological research station at K25. When I was there, the funding was suddenly cut and all the scientists walked out, leaving everything. So we environmental management guys went in there, and there were all of these biological hazard signs: "Do not enter," "Warning, biohazard." We had no idea what was there and we called in reinforcements to clean up the place. We caught one of the new workers we brought in pouring all sorts of stuff down the drain, into the biological treatment system. It killed every germ in the biological treatment system, and a lot of the waste went out untreated into the Clinch River. You would just walk in and see groups doing some crazy experiments with biohazardous materials, and you had to figure out what to do. It was just a crazy time because we were dealing with so many materials and unknowns.

RRJ: Tell us about the mercury that was used there and any incidents with it.

MM: This was over at X10. They were doing breeder reactor coolant experiments. Breeder reactors never got off the ground, but the original idea was to use liquid sodium as a coolant in the breeder reactors. Sodium is hazardous; mix it with water at room temperature and it can cause all sorts of disasters. They said, "Maybe we can use mercury." They bought tons of mercury and put it in the experimental cooling loop. The mercury, being much heavier than sodium, broke a joint in the system. It came to a corner and wouldn't turn. It went right through, and they ended up pumping tons of mercury into the basement of the building. They had to figure out how to get the mercury out of the building. One of the ideas was to knock a hole in the basement wall and drain it out, because it could not be pumped out due to its density and weight. The pumps couldn't handle it.

Workers were splashing around in the mercury while working with cranes that were lifting the mercury into containers to get it out of there. Ironically, years later, an environmental researcher said, "there are abnormally high concentrations of mercury in the fish, other creatures and sediment of Poplar Creek." They said, "No, you're a liar, you're fired, you're crazy, you're causing an alarm. How could we possibly have gotten mercury into Poplar Creek?" They knew all along that they had major spills of mercury there—they were splashing around in it—how could there NOT be major spills of mercury there? To this day, there is a denial. They threatened this environmental scientist with jail!

RRJ: When was that?

MM: I think it was in the early1990s that he was threatened. I had left Oak Ridge by then and read about this incident later. DOE's attitude, when I was there, was that the EPA had no jurisdiction there. It was 10 years after I left that DOE and EPA came to an agreement. The attitude was that national security was more important than the environment. Well, taking care of the environment is national security, in my opinion.

RRJ: Did that attitude change, into the '80s and '90s?

MM: I think it eventually changed, but it took a very long time.

RRJ: What was it like for the workers? Did they know what they were working with?

MM: While I was working the night shift, I did environmental management presentations to the hourly workers, and this was the first time that most of them had heard of it. I think the program was well received by the workers, though. Ironically, the Harvard Medical School had been conducting long-term epidemiological research on the health and mortality of the Oak Ridge workforce, going back to World War II until perhaps until around 2000. The most shocking thing I read was that the average life span for a retiree of that endeavor was 18 months after retirement.

MM: During a criticality incident at the X10 reactor, they lined up hourly workers to go into the manhole to pound on the control rods. They told them to go in for five seconds—that's two good swings with a sledgehammer. That's a lifetime exposure of radiation.

We built a new holding pond for contaminated sludge, based on the hypothetical 100-year flood level. Just a couple of years later, in the spring of 1978, we had a 100-year flood, and it flushed all of the new holding ponds into the Clinch River and Poplar Creek. I think that's when some of the problems with the mercury might have occurred. And then I believe they had a 1,000-year flood a few years later after I left. I hate to think what was flushed out then.

They eventually did get serious about environmental cleanup. I think things are better now. As I said earlier, when I worked there, it was the Wild West. They went from virtually no environmental management in 1970 to some control in the late '70s. But the legacy was appalling. However, Y12 (the weapons plant) was a different animal. The secrecy levels were through the roof. That was where they made the plutonium buttons, the detonating devices for bombs. Y12 is clouded in more secrecy.

On another note not related to Oak Ridge but of interest to your book project, you asked me about Joe Pettit, the former president of Georgia Tech. He was a victim of misuse of radiation when he was a young boy, say in the 1920s or '30s. His father was basically the cause of his death, many years later. His father was a professor of physics in the early days of nuclear physics. Joe Pettit had acne as a young person, and his father treated his acne with radiation. Joe developed a degenerative bone/jaw disease—he had a very deformed jaw (not unlike the Radium Girls). He eventually died of his disease in 1986 while he was president of Georgia Tech. It is pretty well documented that he acknowledged the radiation as the cause of his deformity, saying that people just didn't know as much about it then.

A CONVERSATION WITH YEREVAN (YERO) PETERSON AND LINDA MUSMECI KIMBALL, OXFORD, OHIO—DECEMBER 5, 2010

RRJ: I would like to have a discussion of what happened post–cleanup at Alba Craft. I would also like to discuss what you think this whole phenomenon means in terms of community action in our connection with atomic development. What has happened since 1994–95 when the old machine shop building was razed?

YP: I wonder if you have heard about the Monday morning meetings Linda and I had within the first or second year of the building coming down. We met with the city manager, the mayor, a few managers involved in taking down the Alba Craft operation, a reporter from the *Oxford Press*, as well as the principal of Stewart School that was nearby the site. The principal was interested in the rock crusher, which was crushing up the concrete of the building walls and floors so that it would take less space to transport.

RRJ: I remember that they ground it into dust.

YP: The principal wanted to be there to hear when they would be using the rock crusher so that the children would be indoors, not outside during recess. She took special care letting them know she wanted to be informed so the children would be indoors.

RRJ: How far away was the rock crusher from the school?

LMK: About a city block—perhaps not quite.

YP: Perhaps 300 yards.

LMK: It was impossible for her to keep the kids in all day, because the rock crushing was happening for months, and it would go on all day. Throughout the entire process of the cleanup, we held these weekly Monday morning meetings for many months. In attendance were two Bechtel Corporation supervisors on the job, occasionally Graham Mitchell from the Ohio EPA, and twice a member of the Ohio Department of Health was there. We were trying to get them to hose down the dust from the demolition of the building and the soil. Every time we showed up, someone from Bechtel would hurry and get a hose. We were like patrol officers because we showed up every day. Also at the meetings were the editor of the local newspaper, who wrote stories every week on the ongoing cleanup process, and Graham Mitchell of the Ohio EPA, although he was not at every meeting. Mitchell had the job at EPA of monitoring cleanup of contamination at federal facilities in Ohio.

We were watching them. Every time they tested another grid, they had to show us what they were finding. You know, it was only the community members who were calling them to task, because the EPA and the Ohio Department of Health—who were responsible for compliance with Ohio's environmental laws—would have only looked at the final results; and the final results, had it not been for our oversight, would have shown perhaps not as deep a level being tested. In other words, it would not have shown what was tested along the way. In fact, in answer to your question about the role of "community action in connection with atomic development," I will say that, had it not been for community action, the uranium contamination at this site in this small community would *still* be here today in 2010, awaiting attention and federal funding for cleanup along with the long list of 400+ FUSRAP—Formerly Utilized Sites Remedial Action Program—sites in the country. If you remember, when the DOE conducted its original radiological surveys of the site in July and September 1992, they reported to the city officials and the property owner that it posed no risk to the public health "as long as it was left undisturbed." They also told them that since the site required cleanup of the radiological contamination, it had been added to their list of many sites and would in some years to come be "remediated." City officials and the property owner accepted their assessment of risk and left it at that, keeping the information to themselves. They had no inkling of what the radiological contamination meant, nor did they think of studying the site's history for the pathways the contamination may have taken over the many years. It took a few concerned individuals organized in our community who, once made aware of the designation of the site in March 1993 by fortuitous circumstance, made it all public, raised public concern, investigated the history of the site and the potential pathways of contamination, and pressured authorities to act for a site characterization and a prompt date for cleanup.

I remember very clearly the NRC's [Nuclear Regulatory Commission] ALARA criteria for cleanup, which you mention [in Chapter 8 of] *Romancing the Atom*, was brought up at one of the public meetings.

In the beginning, they were asking the city and the property owners—almost as if in a negotiation process—whether 150 picocuries per liter was an acceptable contamination level to leave behind. Thank goodness for Graham Mitchell at Ohio EPA, who said, "You know, there is such a thing as ALARA." The acronym stands for "as low as reasonably achievable." And the new lowered standard for "safe" exposure to uranium, recently set by the NRC and DOE [Dept. of Energy] and the military when cleaning up their DOD [Dept. of Defense] sites, is 35 picocuries per liter. They were offering us 150 picocuries per liter as a cleanup level. It was as if they were trying to sell us something. "Will you take 150? No, I bid 35. . . ." If we had accepted that, without the knowledge of what was going on elsewhere, what standard for "clean" was set elsewhere, the higher levels of contamination left behind would have been allowed. They were unwilling to dig deeper and look more and get down to a level of 35 picocuries per liter without our insistence.

YP: They were unwilling even to touch the road, except to dig a couple of holes to get a highly contaminated root system from a tree that was in the backyard.

LMK: It cost them another $1M to get it down to at the very least 35, which is 25 picocuries per liter of uranium in the soil higher than natural background levels for our area, which is about 11 at most.

RRJ: Are there stories about the workers?

YP: I can tell you about a woman who lived in a house in Oxford during the 1950s. She was the wife of a worker who was glad to have a job that paid well. They had bought or rented a house, which had several bedrooms that they rented to workers from their town in Kentucky. The workers would sleep in the same rooms using the same beds at different times, taking turns when their time for sleeping came. This lady was washing their sheets and clothes.

I was running into her occasionally during the second year of our involvement with the Alba Craft cleanup, and we began to recognize each other. She confided to me of the many miscarriages she had. She had gone to the same physician I had gone to in Hamilton, OH. And he asked her, "What's happening in Oxford?" He had asked me that question, too, because I had suffered a miscarriage. This was mind-boggling to me because we didn't know how many families had gone through the same thing because of exposure. Hers was because she was handling and washing the workers' clothes.

LMK: I wanted to follow up in the vein you brought up, that there is something broader. In recent years, the incidence of cancer in Oxford—and we're not sure whether you can attribute the higher incidence of cancer to better detection and treatment—but the question has been raised often and publicly: Why is it that in Oxford and the greater area around Oxford, there is such a high incidence of cancer? You must remember that a major nuclear weapons facility—Fernald—is within 30 miles of Oxford. A story appeared within the past five years that reported that the incidence of cancer in southwest Ohio is higher than the national average and is the highest in Ohio. So we're talking about not just radiological but industrial contamination. And Ohio as a whole has a higher incidence than the whole country. But Ohio also has a history of industrial pollution.

One of the former Alba Craft workers said, "Some people died prematurely and some didn't." But those who died, died of radiologically associated illnesses, such as kidney disease and cancer. Eugene (Bill) Albaugh himself died in 1958—one year after uranium milling operations ended at his machine shop in 1957. He was 57 years old. When the contamination story broke in 1993, the few former workers still alive and living in the area told many stories about their experiences working with this strange material with unusual properties and the disposal of waste and the working conditions inside the building where uranium milling operations went on around the clock seven days a week. They were no longer held by the secrecy they were asked to keep about the "important government work" they were doing in the '50s. Some 40 years later, they still had not known that it was elemental uranium-238 they were working with. The total quantity of uranium they machined at Alba Craft was estimated at several hundred tons. U-238 is a heavy metal, likely to be trapped where residues fell, and is an alpha emitter of highly ionizing radiation that cannot penetrate the skin, but, if inhaled or ingested, can penetrate human cells and cause biological damage. And U-238 has a half-life of 4.5 billion years. The risks of exposure are clear—clear enough for most in the medical community to wonder if there is such a thing as a safe or acceptable level of exposure to radioactive materials.

YP: When they were taking down the building in the '90s, they still found uranium dust on the windowsills. That's spooky.

LMK: And the place had been cleaned up twice before by crews from the Fernald facility—cleaned up to the standards in use at the time, like with brooms and mops. But there were still machines and other equipment in there. So just imagine the exposure to the people who worked in that building until it was closed down in the 1990s.

YP: We had a tour of the former Albaugh house with the DOE people. They showed us cupboards, on the second floor, where Albaugh stored his clothing. It was very "hot."

RRJ: And is this where the clothing had the dust on it, which filtered down the inside of the walls to the first floor?

YP: Mr. Albaugh did not change clothing or shoes at the machine shop; he'd kick off his shoes when he sat on his bed on the second floor above the kitchen at home. The uranium dust, we were told, filtered through the floor, through to the kitchen, and into the soil beneath the kitchen. That contaminated soil was dug out and clean soil put beneath the kitchen area. I don't recall hearing how many inches or feet of clean soil were replaced beneath that part of the house. And Albaugh used his garage when he was developing the machines he used for drilling the uranium rods.

RRJ: I heard that when he was negotiating with the government, in 1950 or 1951, he became the chair (or one of the main professors) of industrial arts, where he received a grant for funding to replace a lot of the machinery in the industrial arts laboratory at Miami. And when they brought in the new machinery, he took the old machinery over to the Alba Craft building.

EVJ: Was there any testing on Miami University's campus?

YP: Yes, in the industrial arts building.

LMK: We speculated about where the uranium might be in the city. We put our speculations out there, publicized it, and Miami University requested testing.

YP: But they didn't do it as thoroughly as they might have done. The rooms in the university building had been painted over a number of times since the 1950s.

LMK: Besides, the Alba Craft building itself had been used for more than uranium milling after 1960. And it had been "decontaminated"—cleaned up and cleaned up and cleaned up just as a matter of course by the various occupants over the years. And still, there was enough residual radioactive materials both inside and outside the building as well as off site to exceed current DOE guidelines for "safe" exposure and to warrant a federal cleanup.

RRJ: As you experienced and went through all of this—in the context of this being in a book and being read by people in places other than Oxford—are there other things you would like to share in terms of what this means or says to others?

LMK: I have thought about this. As a member of this community who was personally affected by the radiological contamination, unknown to us, there was an opportunity for me to make a difference, to educate others about what happened here in the '50s and move them to action to press authorities for speedy removal of this contaminated blight in our community, even though I could do nothing to repair the harm it had caused so many. Yet, to me, it's a much bigger issue than what happened in just this one place. There is so much more to be done. We mustn't keep what we've learned to ourselves. We must make people aware of what happened here and in too many other communities. We must teach the history of the nuclear arms race and the risks we took and still take with Americans' health and safety in our country's pursuit of security through superior military power. As former Senator John Glenn put it during one of our lobbying visits in 1994, with nuclear weapons development, we run the risk of "destroying from within what we are trying to protect and defend from without." What I would like to see is nuclear age education, for everybody everywhere, so that none of these lessons are lost.

EVJ: That was Einstein's desire. He wanted to start a new way of thinking.

LMK: Yes.

YP: And once something [like this] happens, people are satisfied and say that it's now behind us.

LMK: So, there seems to be no memory of it, and no one learns the lessons; it's cleaned up, it's fixed. Without formal education, the new generations won't have it; they won't get it. We've learned from Marie Curie, who learned very quickly what she was dealing with. Robert Oppenheimer soon learned. Leo Szilard soon learned . . . and the lessons they learned have to be learned by everyone. Because this thing happened in our community, because we have one of the little pieces in the nuclear production network, and we fixed it . . . that's it?!

The thrust of the "atomic mindset," to use your phrase, is the drive for U.S. security and influence in the world, and what our policymakers perceive to be central to our sense of security is military might/power. The greed of the corporations in the defense industry also drive this thing—there's big money to be made. They are war profiteers, people who make money on the technology and weaponry of war. It has become institutionalized. There is a long history of momentum in this "institutionalized" industry. If everybody had the nuclear-age education about the effects this has had—the environmental footprint it has had on our lives—perhaps we wouldn't be so much in love with the atom.

YP: It's troubling. You might think it is reaching everybody, but it isn't. It strikes me that they said they were testing the soil from the backyards. I never heard what they found in our backyard. I don't believe they did any followup.

LMK: I want to say that the lesson I learned from Oxford is that I could not have done this alone. If we organize and work together, we can make a difference in making positive things happen. But it takes getting off your duff and doing it. The cloak of secrecy that surrounded this work in the '40s and '50s is still there to a great extent, even in our own little community. Nobody knows about it; few have been educated about it. We have to change that scenario and bring knowledge of the nuclear age to students from high school on up. It will take creative, assertive individuals, but I believe they are out there and it must be done. After all, *you're* engaged in this work of educating others about the underside of nuclear weapons development.

Making Up Your Own Mindset: Reading Group Discussion Questions and Classroom Resources

Stories are educational, entertaining, and sometimes shocking, and they help us to understand the world we live in. The stories that comprise *Romancing the Atom* (RTA) are not the whole story of our romance with the atom. In fact, they are just the tip of the iceberg. To further explore the stories—and go beyond just these stories of the atom—this epilogue is designed for you to do some of your own exploration, whether you are a student in a formal school setting or an independent reader who wants to learn and do more. Further, the hope is that you will share your new stories and reflection on *Romancing the Atom* with others in small reading groups or larger venues, such as local community reading initiatives or college reading experience programs.

To this end, the following is broken into three sections. First are questions that arise from the chapters in this book to spur thought and conversation among readers. Second are suggested assignments for teachers to use in high school and college classes. Finally, digital resources are offered for additional reading and research.

DISCUSSION QUESTIONS

- How has RTA affected your views of science, medicine, the military, the government?
- Did Katherine Schaub "die for science," as she suggested in her interview shortly before her death? Of what other people in RTA might we also ask this question?

- Did RTA bring back memories for you of the history of atomic development? Which of these memories most surprised you? That is, were you surprised that you had forgotten them?

- Have you or someone you know been affected directly by moments in atomic development? What do you or they remember from these personal experiences? Have these memories been revived by reading RTA?

- Consider the 1951 photograph of the men sitting in Adirondack chairs observing an atomic exploration in the Marshall Islands. What "story" does that picture tell you? Is it a vision of "romance" with the atom? How?

- Is it possible to "change our modes of thinking" as Albert Einstein suggested? What would this entail? Where might you as an individual begin in such a change of thinking modalities?

- What does RTA bring to mind regarding various environmental issues? Is, for instance, nuclear power a viable alternative to solar and wind power? Why or why not?

- What connections can you make between the "promise" of nuclear energy and, say, the "promise" of deep-water oil exploration? Are there similarities/differences?

- In light of the quotes in support of nuclear power development in Chapter 9 who has the most to gain from investments in nuclear power development? Who might lose?

- If a nuclear power plant is proposed for construction in your regional, what would your reaction be? What actions might you take?

- What specific events immediately come to mind regarding indigenous peoples from your reading of RTA? Why do those moments come so readily to mind?

- What powers do indigenous peoples have with regard to atomic development?

- How would you characterize the conversation that often takes place when members of an indigenous culture express to others outside their group a distrust of atomic science, military uses of atomic energy, or any other aspect of atomic development?

- If you are diagnosed with a potential health problem and the doctors tell you they want to immediately conduct a CT scan, MRI, or X-ray, what might you ask before you agree to the procedure?

- We live in moments of "frenzy" such as the uranium frenzy of the 1950s. In the present day, it might be argued that the rush to embrace digital technologies is a frenzy. Do you think this is the case? Are there similar characteristics in these moments of frenzy, or other historical moments of frenzy (oil, timber, gold, etc.), that we have witnessed?

- What do the events surrounding the Fukushima disaster tell us about the atomic mindset? Do those events reinforce any of the arguments concerning the atomic mindset as put forth in this book?

- How have the 2011 events in Japan affected your perception of the "green enthymeme"? How might you think these events have affected the many people quoted in Chapter 9 who argue for nuclear power prior to 2011?

CLASS ASSIGNMENTS

1. Locate and interview individuals affected by or involved with the atom. For instance, there might be a nuclear power plant nearby with employees who would be glad to do this. Also, professors at a local university who are involved with atomic energy are a good source. And there is a good possibility that there have been people affected by events such as the Alba Craft cleanup who would like to talk about their experiences. This can be done as a group or individual assignment and completed as an oral presentation to the class. NOTE: You should always get written permission from any interviewee, as they are "human subjects."

2. Explore the science of the atom. The elemental makeup of the atom and how scientists have come to understand its secrets are fascinating and very well documented stories of science. Conduct a research project that focuses on one of these stories of the atom and write a formal report or research paper.

3. Study the history of the development of the Nuremberg Code and how it came to be the source for the U.S. human subject code of ethics.

4. Examine the Freedom of Information Act that was used in the 1990s to open the huge collection of post–Cold War documents and other artifacts. Consider what that act provided to the public understanding of the atomic mindset and what we might *not* have known if such a governmental act had not been put into place.

5. Explore the Atoms for Peace movement that began immediately after World War II and that is still functioning as the Federation of American Scientists (see the website for the FAS below).

6. Investigate comic books, cartoons, and popular magazines or books that focus on atomic development. You could also look more deeply into the science kits and games of the 1950s, such as those mentioned in Chapter 2 (see the Oak Ridge University web link below).

7. Create a skit that draws from the stories in *Romancing the Atom*. The skit could be dramatic, historical, or even satirical, like in the spirit of *The Daily Show*, *The Colbert Report*, or *The Onion*.

8. Research and write a story of your own about a moment in the history of atomic and nuclear development.

9. The Fukushima tsunami and Dai-ichi nuclear power plant meltdown have placed the "green enthymeme" in a questionable light. Refer to some of the quotes about the absolute need for nuclear power and do some follow-up research on what these people are saying (or not saying) about nuclear power now.

10. Write a poem about the romance, the mindset, and/or the history of the atomic age.

DIGITAL RESOURCES FOR FURTHER STUDY

http://www.nei.org/newsandevents: The Nuclear Energy Institute Home Page.

http://www.armscontrol.org/: The Arms Control Association Home Page.

www.harveywasserman.com: The home page of activist, journalist, and historian Harvey Wasserman, author of *Solartopia! Our Green-Powered Earth*.

http://www.orau.org/ptp/museumdirectory.htm: The Oak Ridge Associated Universities Museum Directory, which is replete with documents about and artifacts from the history of atomic and nuclear development.

fas@fas.org: The Federation of Atomic Scientists website. Founded in 1945 by scientists who built the first atomic bombs, the Federation of American Scientists (FAS) works to ensure that public policy is informed by intelligent, accurate scientific research.

http://www.ploughshares.ca: A Canadian-based organization devoted to the control and elimination of nuclear weapons.

http://www.bikiniatoll.com/: The official website of the Kili/ Bikini/Ejit Local Government Council and Trust Liaison Jack Niedenthal. This site contains historical documents, information about current initiatives, and stories of the Bikinian people. You can also purchase videos and books about the Bikini Islands and explore travel opportunities to the Marshall Islands.

http://www.sric.org/uranium/: The website of the Southwest Research and Information Center, an organization that works to support the Navajo uranium miners, their families, and their communities.

http://www.lm.doe.gov/default.aspx?id=866: The U.S. Department of Energy website for the Formerly Utilized Sites Remedial Action Program (FUSRAP).

http://www.conelrad.com: A popular history website with a plethora of music, videos, art, and stories by people who worked and lived through the Civil Defense era of the Cold War.

http://www.vce.com and http://www.atomiccentralcom: These two cites are the work of Peter Kuran (author of *How to Photograph an Atomic Bomb*) that contain numerous images and stories from the history of atomic development.

Bibliography

Agricola, Georgius. *De Re Metallica*. Translated by Herbert Clark Hoover and Lou Henry Hoover. New York: Dover Publications, Inc., 1950.

Allardice, Corbin, and Edward R. Trapnell. *The Atomic Energy Commission*. New York: Praeger Publishers, 1974.

Anderson, Rick. "Great Balls of Fire." *Mother Jones*, January 4, 2000.

Argonne National Laboratories. "Post-Remediation Dose Assessment Report." Environmental Assessment Division, Argonne National Laboratories, 1996.

Aristotle. *The Nicomachean Ethics*. Translated by Robert C. Bartlett and Susan D. Collins. Chicago: University of Chicago Press, 2012.

Aristotle. *The Rhetoric and the Poetics of Aristotle*. New York: Random House, 1954.

Assistant to the Secretary of Defense for Nuclear and Chemical and Biological Defense Programs. *Report on Search for Human Radiation Experiement Records, 1944–1994 Vol. I*. Washington, DC: Department of Defense, U.S. Government, U.S. Dept. of Commerce, Technology Administration, National Technical Information Service, 1997.

Associated Press. "Alarms over Radiation." *The Daily Mining Gazette*, October 2010, 5A.

Baker, Nicole, and J. Niedenthal. *Bikini Atoll World Heritage Nomination to UNESCO*. Republic of the Marshall Islands, 2009.

Bellows, Alan. "Undark and the Radium Girls." *DamnInteresting.com*, November 2007.

Bercaw, J., R. Maxfield, L. Stultz, M. Smyser, and M. En. *An Evaluation of the Formerly Utilized Sites Remedial Action Program at Alba Craft Laboratory Oxford, Ohio*. Public Service, Institute of Environmental Sciences, Miami University, Oxford, OH, 1994.

Bergen, Stanley S. "U.S. Radium Corporation, 1917–40." University of Medicine & Dentistry of New Jersey Libraries-Special Collections,

Newark, NJ, October 2001. http://libraries.umdnj.edu/History_of
    _Medicine/USRadiumCorp.html

Bernstein, Jeremy. *Plutonium: A History of the World's Most Dangerous
    Element*. Ithaca, NY: Cornell University Press, 2007.

Bogdanich, Walt. "Medical Group Urges New Rules on Radiation." *New
    York Times*, February 5, 2010.

Bogdanich, Walt. "Radiation Offers New Cures, and Ways to Do Harm."
    New York Times, January 23–24, 2010. http://www.nytimes.com/2010/
    01/24/health/24radiation.html.

Bogdanich, Walt, and Rebecca R. Ruiz. "F.D.A. to Increase Oversight of
    Medical Radiation." *New York Times*, February 10, 2010. http://www
    .nytimes.com/2010/02/10/health/policy/10radiation.html.

Bontrager, Eric. "BLM Authorizes Grand Canyon Uranium Exploration."
    New York Times Energy and Environment, May 2009. http://www
    .nytimes.com/gwire/2009/05/06/06greenwire-blm-authorizes-grand-canyon
    -uranium-exploratio-10572.html?scp=1&sq=BLM%20Authorizes%
    20Grand%20Canyon%20Uranium%20Exploration&st=cse.

Boyer, Paul. *By the Bomb's Early Light: American Thought and Culture at the Dawn
    of the Atomic Age*. Chapel Hill: University of North Carolina Press, 1994.

Boyer, Paul. *Fallout: A Historian Reflects on America's Half-Century Encounter
    with Nuclear Weapons*. Columbus: Ohio State University Press, 1998.

Boyer, Paul. Interview by Robert Johnson. Conversation with the Author.
    Madison, WI, May 10, 2010.

Bradley, David. *No Place to Hide*. Boston: Little, Brown and Company, 1948.

Braffman-Miller, Judith. "Human Experiements: Eileen Welsome/When
    Medicine Went Wrong: How Americans Were Used Illegally as Guinea
    Pigs—Human Medical Experiments on Radiation." *Whale*, March 1995.
    http://www.whale.to/b/welsome_h.html.

Brugge, Doug, Timothy Benally, and Esther Yazzie-Lewis. *The Navajo People
    and Uranium Mining*. Albuquerque: University of New Mexico Press,
    2006.

Bryant, Bunyan, and Paul Mohai. *Race and the Incidence of Environmental
    Hazards: A Time for Discourse*. Boulder, CO: Westview Press, 1992.

Caldicott, Helen. *If You Love This Planet: A Plan to Save the Earth*. New York:
    W.W. Norton and Company, 2009.

Caldicott, Helen. *Nuclear Power Is Not the Answer*. Melbourne: Melbourne
    University Press, 2006.

Center for Biological Diversity. "Bureau of Land Management Defies
    Congressional Uranium Ban, Approves New Exploration North of
    Grand Canyon." *Center for Biological Diversity Newsletter*, May 5, 2009.
    https://www.biologicaldiversity.org/news/press_releases/2009/uranium
    -exploration-05-05-2009.html.

Clark, Claudia. *Radium Girls: Women and Industrial Health Reform, 1910–
    1935*. Chapel Hill: University of North Carolina Press, 1997.

Congressional Subcommittee on Energy and the Environment. *Mill Tailings Dam Break at Church Rock, NM: Oversight Hearing.* Washington, DC: U.S. Government Printing Office, 1979.

Cushman, John H. "Study Sought on All Testing on Humans." *New York Times,* January 10, 1994.

D'Antonio, Michael. *The State Boys Rebellion: The Inspiring True Story of American Eugenics and the Men Who Overcame It.* New York: Simon and Schuster, 2004.

Del Tredici, Robert. *At Work in the Fields of the Bomb.* New York: Harper and Row Publishers, 1987.

Department of Energy. *Documents 48–122.* 1994.

Department of Energy. *DOE Openness: Human Radiation Experiments.* http://www.hss.energy.gov/healthsafety/ohre/roadmap/histories/0475/0475c.html.

Department of Energy. "FUSRAP Brochure." Information, Energy, United States Government.

Eerkens, Jeff W. *The Nuclear Imperative: A Critical Look at the Approaching Energy Crisis.* Edited by V. Gheorghe Adrian. AA Dordrecht: Springer, 2006.

Eichstaedt, Peter. *If You Poison Us: Uranium and Native Americans.* 1st ed. Santa Fe, NM: Red Crane Books, 1994.

Environmental Assessment Division. Postremediation Dose Assessment Report. Environmental Assessment Division, Argonne National Laboratories, Chicago, IL, 1996.

ERC Student Handbook. http://www.hsph.harvard.edu/erc/somehistory.html (accessed October 2009).

Falconer, Bruce. "Uncle Sam's Human Lab Rats." *Mother Jones,* May 18, 2009.

Fankhauser, David. "Independent Report." Oxford, OH: Environmental analysis of Alba Craft, Inc., 1993.

Feiertag, Joe. "Cleanup Timeline, Lack of Funds Frustrate Ohio Officials." *Journal-News,* (Hamilton, OH), March 27, 1993.

Feiertag, Joe. "DOE Can't Foot the Bill for Cleanup: Uranium Contamination Plentiful at Oxford Plant." *Journal-News,* March 26, 1993.

Feiertag, Joe. "DOE Marks 5 Ohio Sites for Cleanup." *Journal-News* (Hamilton, OH), 1 (March 27, 1993): A7.

Feiertag, Joe. "Safety Standards Have Changed Dramatically Over Time." *Journal-News* (Hamilton, OH), March 27, 1993.

Feiertag, Joe, and Kevin Reeve. "Helping Build the Bomb: Area Men Recall World War II Uranium Work." *Journal-News,* April 26, 1993.

Feiertag, Joe, and Kevin Reeve. "Radiation Defense Scant: Only Gloves, Apron Separated Worker, Uranium." *Journal-News,* April 26, 1993.

Frame, Paul W. "Radioactive Curative Devices and Spas." *Oak Ridger Newspaper,* November 5, 1989.

Frankel, Bruce. "Revolt of the Innocents: An Angry Victim Seeks Justice for Schoolmates Once Treated as Guinea Pigs." People.com, May 18, 1998. http://www.people.com/people/archive/article/0,20125312,00.html.

Geiger, H. Jack, David Rush, and David Michaels. *Dead Reckoning: A Critical Review of the Department of Energy's Epidemiologic Research*. Physicians for Social Responsibility. Washington, DC: Physicians Task Force on the Health Risks of Nuclearn Weapons Production, 1992.

Giangreco, D. M. *Hell to Pay*. Annapolis, MD: Naval Institute Press, 2009.

Gill, Brendan. "The Talk of the Town, 'Search'." *The New Yorker*, August 27, 1949.

Gill, Brendan. "The Talk of the Town, In Business." *The New Yorker*, September 12, 1953.

Glenn, John. Letter sent to the Agency for Toxic Substances and Disease Registry, Washington, DC, November 29, 1993.

Goliszek, Andrew. *In the Name of Science: A History of Secret Programs, Medical Research, and Human Experimentation*. New York: St. Martin's Press, 2003.

Gould, Jay M., Benjamin A. Goldman, and Kate Millpointer. *Deadly Deceit: Low-Level Radiation, High-Level Cover-Up*. New York: Four Walls Eight Windows, 1990.

Greger, Debora. *Desert Fathers, Uranium Daughters*. New York: Penguin Books, 1996.

Groves, Leslie. *Now It Can Be Told: The Story of the Manhattan Project*. New York: Da Capo Press, 1962.

Haegele, Greg. "Uranium—'Yellow Monster'—Threatens Grand Canyon." *treehugger.com*, April 1, 2009. www.treehugger.com/files/2009/04/uranium–yellow–monster–threatens–grand–canyon.php (accessed July 19, 2009).

Hanford Health Information Network. *Environmental Health Programs Hanford Health Information Network*. http://www.hanfordproject.com/notes.html

Heidegger, Martin. *Discourse on Thinking*. Translated by John M. Anderson and E. Hans Freund. New York: Harper Torch Books, 1966.

Higgins, Richard. "Haunted by the Science Club, Monetary Offer Can't Erase His Memory." *Boston Globe*, Boston, MA, January 18, 1998. http://pqasb.pqarchiver.com/boston/access/25510269.html?FMT=ABS&date=Jan%2018,%201998.

Hillgartner, S. *Nukespeak*. New York: Penguin Publishing, 1983.

Hodge, Nathan, and Sharon Weinberger. *A Nuclear Family Vacation: Travels in the World of Atomic Weaponry*. New York: Bloomsbury USA, 2008.

Hyman, Stanley Edgar. "Comment, 'Comment'." *The New Yorker*, March 6, 1948.

Johnston, Barbara Rose, ed. *Half-Lives & Half-Truths: Confronting the Radioactive Legacies of the Cold War*. Santa Fe, NM: School for Advanced Research Press, 2007.

Johnston, Barbara Rose, Susan E. Dawson, and Gary E. Madsen. "Uranium Mining and Milling: Navajo Experiences in the American Southwest." In *Half-Lives & Half-Truths: Confronting the Radioactive Legacies of the Cold War*, edited by Barbara Rose Johnstone, 97–116. Santa Fe, NM: School for Advanced Research Press, 2007.

Kass, Nancy. *Informed Consent*. Edited by U.S. Department of Energy. Washington, DC. http://science.energy.gov/bes/research/research-conduct-policies/.

Kelly, Cynthia C. *The Manhattan Project: The Birth of the Atomic Bomb in the Words of Its Creators, Eyewitnesses, and Historians*. New York: Black Dog & Levanthal Publishers, Inc., 2007.

Kimball, Daryl, Lenny Siegel, and Peter Tyler. *Covering the Map: A Survey of Military Pollution Sites in the U.S.* Edited by Maggie Carfield and Emily Green. Washington, DC: Military Toxics Project, 1993.

Kimball, Leslie A., and Kimberly S. Mayne. "Student Voices Not Heard in Alba Craft Lab Issue." *The Miami Student*, Oxford, OH, April 10, 1993.

King Features Syndicate, *Dagwood Splits the Atom!* New York: King Features Syndicate, 1949.

Knoerr, Alvin W., and George P. Lutjen. *Prospecting for Atomic Minerals: How to Look for and Identify Atomic Ores; Stake and Protect a Claim; Evaluate and Sell Your Minerals*. New York: McGraw-Hill, 1955.

Knoerr, Alvin W., and George P. Lutjen. *Prospecting for Atomic Minerals*. Vol. 1. New York: McGraw-Hill, 1955.

Knox, Richard. "Radiation from CT Scans May Raise Cancer Risks." *NPR.org*, December 14, 2009. http://www.npr.org/templates/story/story.php?storyId=121436092.

Kuran, Peter. *How to Photograph an Atomic Bomb*. Santa Clarita, CA: VCE, Inc., 2006.

Lang, Daniel. "A Reporter at Large, 'The Coming Thing'." *The New Yorker*, March 21, 1953.

Lardner, Rex. "The Talk of the Town, The Fourth R." *The New Yorker*, January 19, 1952.

Luccioli, Colleen. *Lawsuit Challenges DOE Uranium Leases in Colo*. August 8, 2008. www.rlch.org/news/lawsuit-challenges-doe-uranium-leases-colo.

Makhijani, Arjun. "A Readiness to Harm: The Health Effects of Nuclear Weapons Complexes." *Arms Control Today*, July/August 2005. http://www.armscontrol.org/act/2005_07-08/Makhijani.

Makhijani, Arjun, and Ellen Kennedy. "Human Radiation Experiments in the United States." *IEER: Science for Democratic Action*. 1. Vol. 3. Prod. Takoma Park, Maryland: Institute for Energy and Environmental Research, Winter 1994.

Markey, Senator Edward J. "Letter to the Nuclear Regulatory Commission." *Congressman Ed Markey*, October 20, 2010. http://markey.house.gov/index.php?option=com_issues&task=view_issue&issue=50&Itemid=152.

Markey, Senator Edward J. "Letter of Transmittal." *American Nuclear Guinea Pigs: Three Decades of Radiation Experiments on U.S. Citizens.* Washington, DC: U.S. Government Printing Office, October 24, 1986.

Marshall Islands Consolidated Legislation. "Constitution of the Marshall Islands." People of the Republic of the Marshall Islands, Archived at the University of the South Pacific, 2005. http://www.paclii.org/mh/legis/consol_act/cotmi363/.

Mason, Bobbie Ann. "Our Far-Flung Correspondents, 'Fallout'." *The New Yorker*, January 10, 2000.

Merchant, Caroline. "Mining the Earth's Womb." In *Philosophy of Technology*, edited by Robert C. Scharff and Val Dusek, 417–428. Malden, MA: Blackwell Publishing, 2003.

Miller, D., and Kevin Bamrough. *Investing in the Great Uranium Bull Market: A Practical Investor's Guide to Uranium Stocks.* Sarasota, FL: Stockinterview Press, 2006.

Moreno, Jonathan D. *Undue Risk: Secret State Experiments on Humans.* New York: Routledge, 2001.

Mullins, Michael. Interview by Robert Johnson, December 16, 2010, Houghton, MI.

Musmeci-Kimball, Linda. Interview by Robert Johnson, October 10, 2009, Oxford, OH.

Naimark, David, Laura Dunn, Ansar Haroun, and Grant Morris. "Informed Consent and Competency: Legal and Ethical Issues." In *Psychiatry for Neurologists*, edited by D. V. Jeste and J. H. Friedman, 391–406. Totowa, NJ: Humana Press Inc., 2005.

National Lead of Ohio (NLO). "Various Top Secret Government Contracts Related to Proposed Program to Develop a Turret Lathe Operation for Machining the SRO Slug." Oxford, OH: Alba Craft, Inc., 1952–57. Archived in the Oxford (Ohio) Citizens for Peace and Justice Archive.

Nealson, Cory. "Studies: Can Uranium Be Safely Mined in Virginia?" *Daily Press*, September 10, 2010, 3.

NEI Nuclear Energy Institute. *News and Events*, June 2010. http://www.nei.org/newsandevents.

Neuzil, Mark, and Bill Kovarik. *Mass Media & Environmental Conflict: America's Green Crusades.* Thousand Oaks, CA: Sage Publications, 1996.

Niedenthal, Jack. *For the Good of Mankind: A History of the People of Bikini and Their Islands.* Honolulu, HI: Bravo Books, 2001.

Niedenthal, Jack. *Op-Ed: The World's Debt to Bikini.* March 11, 2009. http://www.thebulletin.org/web-edition/op-eds/the-worlds-debt-to-bikini.

Niedenthal, Jack. *A Short History of the People of Bikini Atoll*, March 2008. http://www.bikiniatoll.com.

Norrgard, Karen. "Genetics and Society: Human Subjects and Diagnostic Genetic Testing." *Scitable by Nature Education.* Edited by Cheryl

Scacheri. Nature Education, 2008. http://www.nature.com/scitable/topicpage/Human-subjects-and-Diagnostic-Genetic-Testing-720.

Nuclear Energy Institute. "What Policy Makers Are Saying about Nuclear Energy.". http://www.nei.org/resourcesandstats/documentlibrary/safetyandsecurity/reports/what-policymakers-are-saying.

Nurnberg, Clint. "Contamination in Oxford: Old Uranium Lab still Radioactive, Slated for Eventual Clean-Up by DOE." *The Miami Student*, March 26, 1993.

NY Times. "Toward the Bikini Climax." *New York Times*, May 26, 1946.

Office of Environmental Management. *Closing the Circle on the Splitting of the Atom.* Department of Energy, United States Government. Washington, DC: OE, 1995.

Office for the Protection of Research Subjects. *Division of Research and Graduate Studies Office for the Protection of Research Subjects.* http://research.unlv.edu/ORI-HSR/history-ethics.htm.

O'Leary, Hazel, U.S. Secretary of Energy. "Openess." *Gifts of Speech: Women's Speeches from around the World.* December 7, 1993. http://gos.sbc.edu/o/oleary.html.

Olson, Erik, and Elliott Negin. "NRDC Backgrounder: EPA Reverses Ban on Testing Pesticides on Human Subjects." *Natural Resources Defense Council*, November 28, 2001. http://www.nrdc.org/media/pressreleases/011128a.asp.

O'Neill, Dan. *The Firecracker Boys.* New York: St. Martin's Press, 1994.

Oxford, Ohio City Government. "Public Hearing on Alba Craft, Inc. Laboratories Site Oxford, Ohio." Cincinnati, Ohio, April 21, 1993.

Oxford, Ohio Daily Bugle. *Alba Craft's Fallout: A Special Report.* Oxford, OH, 1996.

Peterson, Yero. Interview by Robert Johnson, October 10, 2009, Oxford, OH.

Peterson, Yero, and Linda Musmeci-Kimball. Interview by Robert Johnson, December 6, 2010, Oxford, OH.

Phillips, Kevin. *Democracy and Wealth: A Political History of the American Rich.* New York: Broadway Books, 2002.

Power, Max. *America's Nuclear Wastelands: Politics, Accountability, and Cleanup.* Pullman: Washington State University Press, 2008.

Radiation Exposure. http://hps.org/publicinformation/ate/faqs/radsources.html.

Raymond Berry Microfilms Collection. "Grace Fryer vs. U.S. Radium." Compiled by National Consumers League Collection. Washington, DC: Library of Congress, 1928.

Redberg, Rita F. "Cancer Risks and Radiation Exposure from Computed Tomographic Scans: How Can We Be Sure That the Benefits Outweigh the Risks?" *Archives of Internal Medicine* 169, no. 22 (December 2009): 2049–2050.

Republic of the Marshall Islands Embassy. *History of the Marshall Islands,* 2005. http://www.rmiembassyus.org/History.htm.

Rhodes, Richard. *The Making of the Atomic Bomb.* New York: Simon and Schuster, 1986.

Robinson, George. Interview by Robert Johnson, Houghton, MI, September 5, 2010.

Robinson, Wm. Paul. "Uranium Production and Its Effects on Navajo Communities along the Rio Puerco in Western New Mexico." In *Race and the Incidence of Environmental Hazards: A Time for Discourse,* edited by Bunyan Bryant and Paul Mohai, 153–162. Boulder, CO: Westview Press, 1992.

Ronckers, Cécile M., Charles E. Land, Pieter G. Verduijn, Richard B. Hayes, Marilyn Stovall, and Flora E. van Leeuwn. "Cancer Mortality after Nasopharyngeal Radium Irradiation in the Netherlands: A Cohort Study." *Journal of the National Cancer Institute* 93, no. 13 (July 2001): 1–16.

Rowa, Aenet. *Everything Marshall Islands.* PostNuke. April 6, 2010. http://www.yokwe.net.

"Safe as Mother's Milk: The Hanford Project." http://www.hanfordproject.com/notes.html (accessed April 18, 2012).

Scharff, Robert, and Val Dusek. *Philosophy of Technology.* Malden, MA: Blackwell Publishing, 2003.

Schneider, Keith. "1950 Note Warns about Radiation Test." *New York Times,* December 28, 1993.

Silverstein, Ken. *The Radioactive Boyscout: The Frightening True Story of a Whiz Kid and His Homemade Nuclear Reactor.* New York: Villard Books, 2004.

Simon, Bob. "America's Deep, Dark Secret." *Sixty Minutes,* 2004. New York: CBS News. Television Interview.

Staff, Advisory Committee on Human Radiation Experiments. "The George Washington University Search." *The George Washington University.* June 28, 1994. http://www.gwu.edu/~nsarchiv/radiation.

Staggers, Julie. "Learning to Love the Bomb: Secrecy and Denial in the Atomic City, 1943–1961." Dissertation. West Lafayette, IN: Purdue University, May 2006.

Stair, Ralph M., and George W. Reynolds. *Principles of Information Systems.* 9th ed. Boston, MA: Cengage Learning, 2009.

Strickland, Eliza. *24 Hours at Fukushima—IEEE Spectrum.* November 2011. http://spectrum.ieee.org/energy/nuclear/24-hours-at-fukushima/0.

Subcommittee on Energy Conservation and Power of the Committee on Energy and Commerce, U.S. House of Representatives. *American Nuclear Guinea Pigs: Three Decades of Radiation Experiments on U.S. Citizens.* Washington, DC: U.S. Government Printing Ofice, 1986.

Surbey, Jason. "In Alba Craft's Shadow." Documentary on WMUB Radio, Oxford, OH, September, 14, 1994.

Szczygiel, Tony. "UB Center for Clinical Ethics and Humanities in Health Care/Bioethics Law in New York State: Mary E. Schloendorff, Appellant, v. The Society of the New York Hospital, Respondent." *University at Buffalo.* http://wings.buffalo.edu/faculty/research/bioethics/schloen0.html.

Time in Partnership with CNN. "Medicine: Poison Paintbrush." *Time*, June 4, 1928.

Time in Partnership with CNN. "Medicine: Radioactive!" *Time*, February 3, 1958.

Time in Partnership with CNN. "Medicine: Radium Hangovers." *Time*, November 10, 1958.

Time in Partnership with CNN. "Medicine: Radium Poisoning." *Time*, April 1, 1929.

Time in Partnership with CNN. "Medicine: Radium Women." *Time*, August 11, 1930.

Time in Partnership with CNN. "NA." *TIME*, November 6, 1933.

Time in Partnership with CNN. "Science: Death of Radium Painter." *Time*, November 26, 1928.

Time in Partnership with CNN. "Theater: New Musical in Manhattan, February 23, 1953." *Time*, February 23, 1953.

Udell, Gilman G. (compiler). 79th Congress. *Atomic Energy Act of 1946 and Amendments.* Public Law 85, U.S. House of Representatives. Washington, DC: U.S. Government Printing Office, 1971.

United Nuclear. http://www.unitednuclear.com.

United States Congress. House Committee on Interior and Insular Affairs, Subcommittee on Energy and the Environment. Transcript of House of Representatives hearings on the Mill Tailings Dam Break at Church Rock, New Mexico. Washington, DC: U.S. Government Printing Office, 1979.

U.S. Atomic Energy Commission. *Prospecting for Uranium.* Washington, DC: Author, 1951.

U.S. Department of Energy Formerly Utilized Sites Remedial Action Program. *Radiological Survey Results for Neighboring Properties of the Alba Craft Site Oxford, Ohio.* Compilation, Department of Energy, Washington, DC, 1994.

U.S. Radium Corporation East Orange, NJ. "History." *Revised Work Plan.* Vol. 1. Compiled by University Archives/University of Medicine and Dentistry of New Jersey, Newark, NJ: U.S. Radium, 2001.

Walle, Douglas. "The Dirty Little Secrets of the Atomic Age." *Newsweek's Education Site.* December 20, 1993. http://www.newsweek.com/id/125159.

Wasserman, Harvey. Interview by Robert Johnson, December 5, 2010, Bexley, OH.

Wasserman, Harvey, and Norman Solomon. *Killing Our Own: The Disaster of America's Experience with Atomic Radiation.* New York: Delacorte Press, 1982.

Weart, Spencer R. *Nuclear Fear: A History of Images.* Cambridge, MA: Harvard University Press, 1988.

Welsome, Eileen. *The Plutonium Files: America's Secret Medical Experiments in the Cold War.* New York: Random House, 1999.

Wilkins, David E. *Documents of Native American Political Development 1500s to 1933.* New York: Oxford University Press, 2009.

Wiltbank, Ashley H. "Informed Consent and Physician Inexperience: A Prescription for Liability?" Edited by Sean Mazorol. *Willamette Law Review* 42, no. 3 (Summer 2006): 563–566.

Winkler, Allan M. *Life Under a Cloud: American Anxiety about the Atom.* Oxford, UK: Oxford University Press, 1993.

Wright, Irene. "Cleanup Unites Oxford: Uranium Contamination Draws City, Group Together." *Cincinnati Enquirer,* Cincinnati, OH, March 25, 1993, B5.

Wright, Linda. "DOE Reports Oxford Site for Clean-Up." *The Oxford Press,* March 25, 1993.

Wright, Linda. "Owner Planning Move from Contaminated Site." *Journal-News,* March 25, 1993.

Wright, Linda. "Oxford Group Seeks Study." *Journal-News,* March 24, 1993, 40.

Wright, Linda. "Residents Debate Uranium Site: Oxford Citizens Agree Hazard Notification Arrived Too Late." *Journal-News,* March 25, 1993.

Yazzie-Lewis, Esther, and Jim Zion. "Leetso, the Powerful Yellow Monster." In *The Navajo People and Uranium Mining,* edited by Doug Brugge, Timothy Benally, and Esther Yazzie-Lewis, 1–10. Albuquerque: University of New Mexico Press, 2006.

Zoellner, Tom. *Uranium: War, Energy, and the Rock That Shaped the World.* New York: Penguin Group, 2009.

Index

About the Author

Robert R. Johnson is professor of rhetoric and technical communication at Michigan Technological University. An active writer over the past twenty years, he has published in many major journals in his respective fields of science and technology studies, technical communication, rhetoric, and education. His entire career has been devoted to the study of science and technology and how humans define, interact with, and are often controlled by these elemental aspects of existence. His award-winning book (National Council of Teachers of English "Best Book of 1999" in Technical and Scientific Communication), *User-Centered Technology: A Rhetorical Theory for Computers and Other Mundane Artifacts* (1998), delves into the problems people often have with using and being used by technology.

For further information, see www.romancingtheatom.com.